The Soma

How our genes really work and why that changes everything!

Robyn Lindley

© Copyright 2009
First Edition, March 2010
Robyn A. Lindley
CYO ERADE Village Foundation
11 ERADE Drive
Piara Waters WA 6155 Australia

ISBN: 1451525648
ISBN-13: 9781451525649

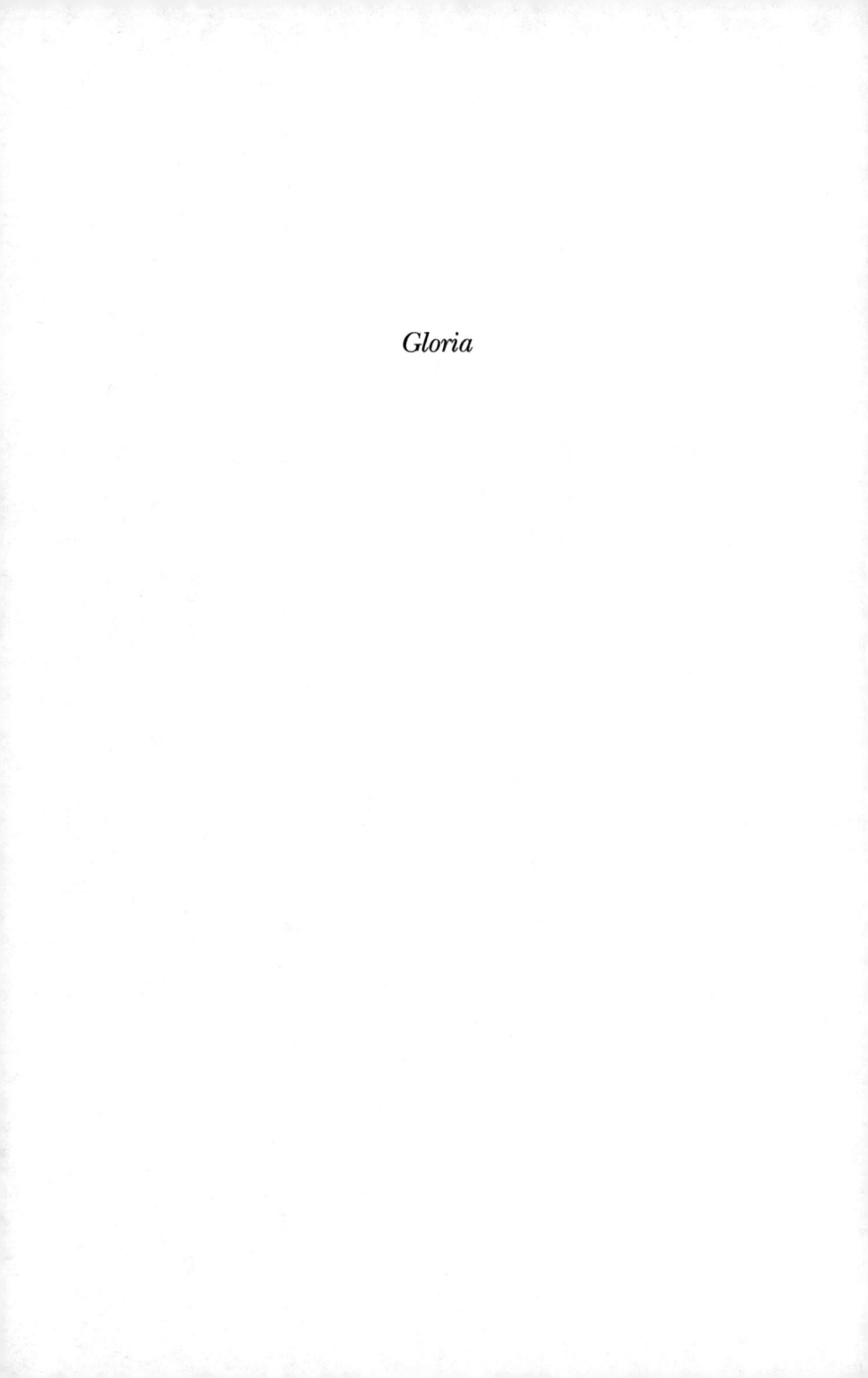

Gloria

CONTENTS

Foreword ix

1. Early ideas on acquired inheritance 1

2. Some heritable genetic transfer effects 21
   My son looks like my wife's ex-husband....................22
   Pekin ducklings...29
   A pink iridescent glo-chihuahu.................................31
   Gene therapy...35
   Gene transfer and adaptive mutation in microbes 38
   GM crops, super weeds and super bugs......................44

3. Sperm mediated gene transfer effects........................49
   Genetic hitchhikers on the surface of sperm..................50
   Smoking can cause sperm damage 52
   Health risk and evolutionary consequences.................53

4. The fluid nature of our hereditary material.............55
   Why DNA alone doesn't tell us much.........................56
   Making RNA from DNA 60
   Making protein from mRNA......................................62
   The universe of RNAs 68
   The cell factory..71
   Updating our DNA 77
   A missing link?...83
   Rearranging our genes 84
   Our genes as a 'living code'..85

5. Acquired inheritance effects in the immune system 89

Early experiments 90
The immune system in action 95
Recognition and response 97
Creating new immune system genes 99
Acquired inheritance in koalas infected
  with an AIDS-like retrovirus 102
The immune system is not blind 104

6. Epigenetics: the footprints on our DNA 107

The same DNA—but oh so different! 109
You are what your parents eat—sometimes! 113
More mice tails 116
Epigenetic inheritance of disease 118
The mule and the hinny 120
Why the canary builds a nest in springtime 121

7. Memories in molecules 127

How good was your childhood? 129
Synaptic plasticity and learning to swim 134
The next Holy Grail 137
Short term and long term memory storage 143

8. Lamarck's night of doubt 149

Why are some fossils missing in the evolutionary record? ...150
So many miraculous accidents in design 154
When does information become a 'fact'? 158
Resurrecting Lamarck 162
The Emerging Evolutionary Paradigm 164

9. Evolutionary learning logic 167

Evolutionary learning logic in the immune system 170
Learning to smell the cheese 171
Genes learning to take on a new role 173
A Lamarckian view of learning language 175
Learning and instinct 180

Learning and intelligence 181
Lamarckism and societal change 182
What makes us human? 187

10. Conscious evolution......................................197
Evolutionary thought and conscious evolution 197
Genetic Responsibility 200
Introducing HomoSap2...202
Technology and conscious evolution.....................207
Genetic discrimination 211
So where to from here? 217

Acknowledgements 219

Appendix...221

Glossary..223

Index...227

References and notes...235

FOREWORD

Evolutionary theory is rapidly approaching a new crisis point. In the context of the new genetics, it is becoming evident that an evolutionary view based on random mutations and survival of the fittest has run its course. While the forces of natural selection do play an important role in evolution, they are only a part of the emerging picture. The massive volumes of molecular and genetic data harvested over the last decade or so are telling us that some strangely Lamarckian phenomena are also playing a crucial role in our evolution.

The new genetics is demanding that we form a more sophisticated world view, based on a synthesis of the Darwinian idea of natural selection, and the distinctly Lamarckian idea that changes acquired during our lifetime are inherited.

This book considers the role of Lamarckian inheritance effects in the context of the new genetics and how that is forcing us to change our views on evolution and our lives. I wanted to write about this because I want people to understand that there is now extant scientific evidence suggesting that nature has evolved a number of molecular mechanisms for us to consciously direct our own evolutionary future. We ignore this at our peril.

It begins with a brief introduction to the history of our ideas on acquired inheritance. It describes how our genes really work, and it brings to life the current plethora of scientific evidence suggesting that every moment of our existence we can potentially alter the makeup of our heritable material.

I have also re-opened some of the thornier debates associated with the idea of Lamarckian inheritance. In the context of the new genetics, these fundamentally alter our understanding about ourselves, and our survival as a species as we reach a new pivotal point punctuating our own evolution.

* * *

Having written *Lamarck's Signature* with Ted Steele and Bob Blanden in 1997,[1] it occurred to me that there is a need to write about the far bigger picture now emerging from the new genetics. In *Lamarck's Signature*, we focused on the immune system and the complex bio-molecular pathways used to enable us to mount an effective immunological response to almost any foreign pathogen. Once an effective antibody is produced, some of the new DNA sequence changes responsible for encoding it may be copied into our DNA. The updated information is then inherited by subsequent generations. However, even after biologists recognised this, it was still widely believed that the process of updating the variable regions of antibody genes did not apply to other families of genes, and so the wider Lamarckian implications were largely ignored.

To understand the science involved in the processes that result in updating some of the variable regions of antibody genes, the reader was presented with facts as they emerged from a deeply historical and scientific perspective. Along the way, the reader was required to grasp a number of fairly sophisticated concepts using specialist terminology. The feedback suggested that this was 'pretty heavy going' for most readers new to the field.

Since we wrote *Lamarck's Signature* much more genetic sequence information has become available, thanks to several large genome sequencing projects, and some dedicated genomic analyses. Armed with a whole new generation of fast sequencing technologies for analysing genetic data, we are now beginning to understand that our genes hold many more Lamarckian secrets. We are beginning to understand the sheer complexity of the genetic machinery at work every moment of our life in terms of the highly regulated pathways, and the molecular machinery interacting with our genes. A three-dimensional picture of a highly re-configurable genome responding to environmental feedback is what is emerging.

What we are also discovering is that many aspects of our own lives can have a lasting impact on our evolution as a species. What we eat, what diseases we develop, how we behave to our young and to each other, and what we do can all alter our genome. Even your cigarettes or sunbaking can leave a specific genetic marker behind in your genes. So there we have it, Lamarckian forces are at work, and we are only just beginning to understand how this occurs, and what it means.

For those that wish to access the original scientific papers and other works used as a primary source, references and notes are included. A Glossary of the key scientific terms is also included. I have limited the number of scientific terms used so that the science itself is more accessible for general readers. For readers not familiar with the field, a single diagram is included in the Appendix, Figure 1. While the science underlying many of the concepts we rely on is complex, this is the only reference map necessary to begin the journey of discovering just how fluid our genes are.

Towards the end of the book, I have also allowed myself to be drawn into a world of consciously directed evolution that transcends the limitations of current evolutionary thought. By drawing on our understanding of the fluid nature of our heritable material and the choices presented to us by gene technologies, we are able to choose our future evolution as a species. One of the main implications of accepting a Lamarckian world view is that each of us has the ability to play a role in directing our evolutionary future. This is a world that is much deeper, and it demands a new sense of self.

The journey of transcendence has also taken me to the edge of Alice's Looking Glass where I am confronted with a new self-image. The new self-image with which I am confronted is one of freedom: acceptance of the elements of acquired inheritance sets me free from the genetic determinism with which I had been indoctrinated. I become more open to consciously conceiving and directing my own evolution, and that of my children. As I pass into this world, my old 'sense of self' is not altered. It is shattered!

* * *

1. EARLY IDEAS ON ACQUIRED INHERITANCE

"... The offspring resembles its parent because the particles of the semem come from every part of the body."

Hippocrates, VII, 471-75

The idea of 'pangenesis' is ancient. It is a theory based on the idea of an hereditary transmission process in which all parts of an organism contribute to the characteristics of the next generation. While variations of the ancient hypothesis of pangenesis have continued to reappear to the present, the mechanisms of adaptation have been debated by philosophers and scientists for more than two millennia. But what brought the ancients to this world view?

Fragments of the idea of pangenesis can be found before Socrates. However, one of the earliest detailed recorded proponents of the inheritance of acquired characters was the father of modern day medicine, Hippocrates of Cos II (460-370 BC). Hippocrates has described how every part of the body produces its own particles that circulate throughout the body, and that these are carried to the testicles.

Based on his observations of the Macrocephali people living near the Euxine Sea with elongated heads, Hippocrates wrote that the elongated heads were first produced artificially by binding the heads of babies to produce long heads. After a time it became an inherited characteristic.[2] According to Hippocrates, once the practice was abandoned the long head shape faded after a number of generations. While most scientists remain sceptical about the acquired inheritance phenomenon described by Hippocrates, there are many records describing this practice amongst other peoples. The practice of moulding the head of infants has been reported amongst inhabitants of some provinces of Germany, the Belgians, the Gauls, some Italians, the Turks, several indigenous peoples of North America, and the Peruvians.[3] While collectively, these historical works do not provide any scientific evidence, they appear to provide unanimous testimony of many eye witness accounts of the changes caused by the practice resulting in infants being born with elongated heads.

There is also some scattered evidence that similar views had been held before Hippocrates.[4] The ancients believed that there are many other characters of an organism that can vary significantly in one's lifetime, and that these changes may be inherited by offspring. Certain muscles are built up, bones respond to different postures, and organs respond to the demands put on them. From this, the idea emerged that some sort of hereditary information relating to changes in a single character were heritable. That is, the ancients believed first, that there is a point-to-point correspondence between the physical characteristics, and those inherited by offspring; and second, that changes in form are heritable.

Although Aristotle did not agree with the pangenetical view that hereditary transmission of acquired characters worked via 'particles', he was not able to argue successfully against it. He reasoned that many heredity traits developed after reproduction, and that they could hardly be carried to the germ line and passed on in gemmules. He simply found it unimaginable that there existed particles for voice, hair or temperament but could not prove or disprove it using the limited observations available to him.

The idea of pangenesis continued to reappear in the literature and was widely accepted until the scientific revolution of the seventeenth century. There was a leap in scientific progress thanks to the importance placed on experiment and reductionist thinking. To a large extent the scientific revolution was because of the adoption of a new mechanistic view of nature. It displaced the older vitalistic animist explanations for natural phenomena with mechanical explanations based on what was referred to as natural law: it was a time when the communal values of science and the scientific method became highly valued.

As a consequence, the development of scientific thought was protected and nurtured by new distinguished societies such as the Royal Society of London established in 1660, and the French *Academie des Sciences* established in 1666. The Royal Society of London adopted a Latin mantra for its members, *Nullius in verba*. Translated, it means that no man's word shall be final, or that no idea shall be presented without doubt. So when the newly invented microscope failed to find the 'particles' responsible for defining characters, this spirit of questioning meant that the idea of pangenesis was again in dispute.

The ensuing scientific debates were based on speculation and limited observations. Yet the idea of pangenesis persisted up until 1809 when Jean-Baptiste de Lamarck provided his account of the inheritance of acquired characters. In 1809 he published *Philosophie Zoologique.*[5] It was a revolutionary work that described the first mechanistic view of how variation is acquired and maintained in an evolutionary context. Lamarck had decided that once the difficult step of 'spontaneous' generation of the first life forms was made, that nature had the means to produce all other forms of life from it. He believed that the primary cause was the result of two main forces.[6] The first was what he described as a progressive force pulling life from its simplest forms, to the more complex forms up the natural 'animal scale'. The second, was the adaptive force which fashioned life's adaptive complexity and which Lamarck believed was due to 'greater susceptibility to modification by the environment'.

Lamarck was confident in making this extraordinary claim as he believed that he had correctly identified the primary cause which tends to make organization more complex over time. He believed that he understood how the environment altered specific forms by modifying their function, and he was convinced that organic change followed a change in habit. He reasoned that two principles were responsible for the gradually increasing adaptive complexity he observed in the animal kingdom.

The first principle that he relied on to explain increasing adaptive complexity was 'the use and disuse of parts'. He proposed that characteristics needed in one's lifetime were passed on to offspring, and that those that weren't needed became diminished. That is, those parts that are used are strengthened and those that are not are reduced in size until eventually they may disappear. The long neck of a giraffe, the weakened moles' eyes, and absent snakes' legs are just some of the examples he used to support his theory. In the preface of *Philosophie Zoologique* he wrote,

> *"First, a number of known facts proves that the continued use of any organ leads to its development, strengthens it and even enlarges it, while permanent disuse of any organ is injurious to its development, causes it to deteriorate and ultimately disappear if the disuse continues for a long period through successive generations."*

Lamarck's ideas on the 'use and disuse of parts' relied heavily on references to bodily fluids. He considered the role of environmental factors in triggering a fluidic stimulus-response interaction as important in producing heritable organic changes. To introduce this principle in the preface of *Philosophie Zoologique* he wrote,

> *"Second, when reflecting on the power of the movement of fluids in the very supple parts that contain them, I soon became convinced that as this movement is accelerated, the fluids modify the cellular tissue in which they move, open passages in them, form various canals, and finally create different organs, according to the state of the organisation in which they are placed."*

He argued that together, these two principles of 'use and disuse' and the associated fluidic stimulus-response, give rise to the extraordinary diversity of forms that we observe in the animal kingdom.

In substantiating his theory, Lamarck was strongly influenced by the work of the renowned French naturalist and zoologist Georges Cuvier. In the 1790s

Cuvier's pioneering anatomical research had produced the first highly ordered classifications of living things based on anatomical differences. Lamarck used Cuvier's ordered classifications to argue that if all of 'nature's productions' were arranged linearly from the simplest forms to the most complex, then all life could be traced back to a single form. This idea continued to dominate Lamarck's thinking for many years. He used it as a springboard to provide a causal explanation for his ideas on what he called the 'law of progressive development'. It provided an important cornerstone for much of his lifetime of work on use and disuse that he believed was responsible for generating adaptation.

A key process that Lamarck relied on to explain his concept of organic development was increasing 'composition' based on his own observations of the development of 'internal' characters such as the circulatory system or central nervous system. He noted that these grew in complexity in organisation over time. These are the characters that Cuvier used to develop zoological classification tables. Lamarck called these the 'power of life' and he believed that the external and internal characters were the result of particular environmental influences. But what did Lamarck mean by the term 'power of life'?

Without the scientific tools we take for granted, Lamarck was unable to reduce life to the biochemical and molecular terms we rely on today. Instead, he described life in mechanistic terms. His explanation was that the main difference between living forms and inanimate objects was that life relied on sensory inputs from the environment. He described how life developed over time in terms of the behaviour of 'fluids' in motion, heat and electricity. He described in great detail how these were responsible for making life possible, and for creating diversity. In the many examples describing the importance of fluids, Lamarck emphasised the importance of environmental stimulation as a vital force for activating fluid flows to animate an organism. He observed that "vital movements are never transmitted but always stimulated" by some environmental factor. In the Introduction to Part II of his *Philosophie Zoologique*, he declared it obvious that these forces have acted to modify organs such as the digestive tract over time. But not everyone was convinced.

Another concept that Lamarck paid particular attention to was the nervous stimulation required to produce movement which he called 'irritability', and feelings that are associated with fluid flow. He concluded that these operate independently. His observations led him to conclude that the nervous system in simple animals is only used to produce muscular movement, and that they therefore have no feelings. In each case ganglia with their tiny 'threads' are involved. In regard to the feelings developed by more complex animals, Lamarck noted that the feelings or sensations were associated with a fluid flow that starts from the point affected.

In putting these ideas forward, Lamarck was in every sense the father of modern evolutionary thought. At the French school Transformationism, established by Lamarck in Paris, an anatomist Robert Edmund Grant wrote an anonymous paper in 1826 praising Lamarck, and making the first known reference to the word 'evolved' with the meaning we now attach to it. However, Lamarck's theories were rejected by most of his peers on the grounds that they were scientifically indefensible. From an historical perspective, it was his detailed descriptions of the mode of transmission of acquired inheritance effects to the next generation that attracted the harshest criticism.

Lamarck died, blind, almost friendless, and without enough money to pay for his funeral in 1829. Yet despite the ridicule and criticism endured by Lamarck in the last decades of his life, the idea of the inheritance of acquired characteristics was almost universally accepted for most of the remainder of the nineteenth century.

Around ten years after Lamarck's death, English naturalist Charles Darwin set out to explain his ideas on how evolutionary change occurs over millions of years. Darwin's developing views on evolution were given greater credence in the mid nineteenth Century due to the work of another biologist Alfred Wallace. Both Charles Darwin and Alfred Wallace were highly respected scientists of the day, and they independently proposed similar views on how evolution occurs. On the eighteenth of June in 1858, Darwin received a copy of a paper from Alfred Wallace, who had been working in the island of Ternate in Indonesia collecting specimens to sell to naturalists, including Darwin. His paper was entitled *On the Tendency of Varieties to Depart Indefinitely From the Original Type.* He asked Darwin to pass the paper on to Charles Lyell at the Linnean Society for possible publication. The process of evolution described by Wallace was surprisingly similar to what Darwin had been working on. This created a moral dilemma for Darwin, who knew that credit would go to the person who published first. However, he forwarded the manuscript to Charles Lyell at the Linnean Society.

It was serendipitous that Darwin had sent a copy of his own theory to botanist Joseph Hooker around twelve years earlier. In the end, both papers were read at the Linnean Society on the first of July, 1858.

The following year, two decades after starting work in the area, Darwin published his most famous work, *The Origin of Species* (1859). In the very first chapter, Darwin makes his views on Lamarckian inheritance apparent, writing, "Many facts clearly show how eminently susceptible the reproductive system is to very slight changes in the surrounding conditions". Chapter 1 also included a section on the *Effects of Habit and of the Use or Disuse of Parts* in which he provided his description of the ancient idea of pangenesis. He wrote,

> *"... I find in the domestic duck that the bones of the wing weigh less and the bones of the leg more in proportion to the whole skeleton, than do the same bones in the wild-duck; and this change may be safely attributed to the domestic duck flying much less, and walking more, than its wild parents."*

To explain his ideas, he described changes in many domesticated animals compared to ones of the same species still living in the wild. For example, he noted the increased size of udders in cows and goats that were regularly milked. He noted the tendency for the ears of domesticated animals to droop due to the disuse of these muscles. At the end of the first chapter of *The Origin of Species* Darwin concluded,

> *"Changed conditions of life are of the highest importance in causing variability, both by acting directly on the organism, and indirectly by affecting*

> *the reproductive system... The greater or less force of inheritance and reversion, determine whether variations shall endure,... perhaps a great, effect may be attributed to the increased use or disuse of parts."*

Darwin was aware that a theory based on the forces of natural selection alone could not explain how many of the physical features he examined arose. In regard to the evolutionary origin of the human eye he devoted a long section to explain how a simple eye spot consisting of an optic nerve surrounded by pigment cells, with no lens and covered in a translucent skin, could develop into an organ as perfect as an eagle's eye. He described the mode of transition as a series of slight structural modifications that result in many intermediate forms. He concludes in Chapter 6 of *The Origin of Species* that the difficulty of believing that something as perfect and complex as an eye could be formed by natural selection alone, should not be considered as subversive of his theory. Darwin was also very puzzled about how a natural instinct arose in the first place. In fact he saw these and the other examples he cited as potentially fatal to a theory based on the forces of natural selection alone.

While Darwin carefully described the forces of natural selection as being a dominant evolutionary force, he went to great lengths to acknowledge that acquired inheritance effects also played an important role. Darwin put forward his own ideas on the mechanisms for the inheritance of acquired characters. He did not see these as being in conflict with the processes of natural selection. However, his opponents tried to erode the value of his work by associating it with that of Lamarck. Darwin would have been acutely aware of how

Lamarck's ideas were ridiculed, and he tried to distance himself from them. It didn't help when the influential Scottish geologist Charles Lyell referred to Darwin's work as a modified version of Lamarck's theory.[7]

Darwin's solution to the dilemma was to essentially ignore Lamarck's work—for instance, making no direct reference to Lamarck in *The Origin of Species*. He does however, acknowledge Lamarck in his introductory historical sketch. Buried in the *Recapitulation and Conclusion* of *The Origin of Species* Darwin also makes reference to Lamarck as the person who 'first called attention to this subject.' By 'this subject' Darwin meant the idea that all species probably descended from the same parents. Yet he continued to develop his own ideas about acquired inheritance effects.

In 1868, Darwin sketched out his ideas on pangenesis in greater detail in his book *Variation in Plants and Animals Under Domestication*. In it he introduced many examples to demonstrate the inheritance of acquired characters. He also focused on the causes of variation, and he put forward what he called his *Provisional Hypothesis of Pangenesis*, without acknowledging predecessors. Unlike much of Darwin's writing, this work did not attract much commentary either way.

Darwin even put forward a theory of the mechanism to explain how he thought acquired inheritance occurred based on minute 'gemmules' that are a characteristic feature of all living cells. He speculated that a somatic cell would throw off what be called 'gemmules' in response to environmental stimulation. He believed that these gemmules were altered through the modification of cells. Most remarkably, he postulated that somehow these gemmules were passed into the gonads where they could then be inherited by offspring.[8] The critics argued that this was little more than a synthesis of the pangenetical concepts that had been circulating since the time of Hippocrates of Cos II (460-370 BC). However, in putting forward these concepts, Darwin believed that he had articulated the first molecular mechanisms that could give rise to acquired inheritance effects in the modern scientific era.

A key point of difference between the original idea of pangenesis and Darwin's theory is that he had evoked a separation between the hereditary material ('gemmules') and the rest of the body of an organism.

If we compare a translation of Lamarck's *Philosophie Zoologique* with Darwin's ideas on pangenesis the first thing that is obvious is that they were each written by intelligent and articulate individuals who were proposing the same ideas. Both Darwin and Lamarck described the first comprehensive systems and causal explanations for evolution.[9] Both relied on the idea that environmental stimulation elicits a response from an organism, and the idea that if a response is evoked frequently it might become an acquired inheritance effect. However, they each proposed a different mechanical explanation. In the case of Lamarck, his lengthy explanations involved a flow of fluids that gave rise to changes. Darwin relied on the pangenetical idea of tiny gemmules that circulate through the body in response to environmental signals. Yet twentieth century interpretations of their works have put them into direct opposition with one another. Why have we accepted misrepresentations of their respective works?

To ensure that his own ideas would endure, Darwin was careful to build and maintain a scientifically

and politically influential phalanx of supporters—
something that Lamarck failed to do. He praised the
work of Professor Thomas Huxley who later became
known as 'Darwin's Bulldog' for supporting Darwin's
ideas. He carefully nurtured his relationships with
John Lubbock who made some key contributions
to *The Origin of Species*, and Alfred Wallace, who in
apparently coming up with the same theory had
forced Darwin to rush the completion of *The Origin of
Species*. He also became a close friend and mentor of
the renowned geologist Charles Lyell. He continued
to nurture his relationship with Fritz Muller, an
extraordinarily gifted German naturalist who lived in
Brazil and supported Darwin's theory of evolution.
There were also several other pre-eminent scientists
who helped to build a wide acceptance of Darwin's
ideas on evolution.

In the 1860s and just a few years after
Darwin published *The Origin of Species*, Austrian
anthropologist and monk Gregor Mendel was
busy conducting the first experiments to trace the
physical characteristics of successive generations of
plants. In a short monograph, *Experiments with Plant
Hybrids*, he provided the first description of how
genetic information is passed on from generation
to generation. Although Mendel conducted his
experiments in the 1860s, his work remained
relatively unknown until well after his death in
1884. Around 1900 a number of reports of his work
were published as the science of genetics began to
develop. Mendel later became known as the father
of modern genetics, and is now widely credited with
the inception of the idea of genetic exchange.

In one experiment Mendel cross-pollinated smooth round yellow peas (YY) with wrinkly green peas (GG). All of the first generation of peas were yellow and round like the original yellow peas. They showed genetic dominance effects of the colour yellow over green, with the round smooth shape being dominant over wrinkled peas. Genes responsible for the green colour and wrinkled appearance of the green peas are referred to as recessive—even if they are present in offspring. They will be hidden whenever a Y gene is also present. He went on to show that in subsequent generations, only those peas that inherited two of the genes for green (GG) were green in colour. This is much like the effects of recessive genes for red hair or blue eyes.

According to the Mendelian paradigm, genes have fixed positions on a chromosome, and two copies of each gene are inherited in equal proportion, but only one is expressed. Not all factors are dominant, and some are recessive. The two central ideas it embodies are: the concept of particulate inheritance and the unchanging nature of the inherited particles. The Mendelian concept of the gene as an unchanged factor transmitted down the generations is a foundational concept of twentieth century genetics. It is not supportive of the idea of acquired inheritance.

Ultimately it was German biologist August Weismann's work that delivered the final blow to accepted nineteenth century ideas about acquired inheritance. Weismann's *"theory of the germplasm"* (1893), presented a new hypothesis that, for the first time, was based on the idea of two fundamentally different types of cells, and two types of cell division. One group are the somatic cells (*somatoplasm*), or ordinary body cells that reproduce through a cell division process called mitosis. The second group, are the reproductive germline cells (*germplasm*) that reproduce to produce the next offspring. When you grow skin to heal a graze, you are using somatic cells. When you conceive a child, you are using germline cells (or sex cells).

Weismann's hypothesis was that the somatic and germline information did not mix. This idea became known as 'Weismann's barrier'. It has been an influential and highly effective scientific construct: it was designed to promote the significance of the forces of natural selection in evolution at a time when the science of genetics was being established.

Decades later, Weismann's view became linked to Mendel's laws of genetic inheritance, forming the foundations upon which the neo-Darwinian world view was built. Together these formed the basis for the more comprehensive neo-Darwinian view of hereditary transmission that emerged as we learnt more about our genetic material.

The current form of neo-Darwinism arose sometime after the 1920s. The ideas of natural selection, random point mutations and the concept of Weismann's barrier were brought together to reinforce each other. Random point mutations were relied upon to explain how new genetic variations arose. By uniting these new genetic concepts with Darwin's idea of natural selection, the neo-Darwinian world view became widely accepted.

For the remainder of the twentieth century the neo-Darwinians demanded the total exclusion of Lamarckian inheritance effects. The idea of 'survival of the fittest' became the most dominant scientific dogma of the twentieth century. The notion that characteristics could be developed during a lifetime and then passed on was seen as absurd. Lamarck's ideas were ridiculed in textbooks, and his ideas used as one of the most disreputable examples of an erroneous theory. As a consequence, the idea of the survival of the fittest has remained a powerful influence on our thinking until now.

Yet there have always been Lamarckians. They were seen as heretics—or worse. This meant that anyone contemplating conducting genetic experiments on acquired inheritance effects would need to argue against the strong current of scientific belief. The literature is littered with suicide, admonishment and sackings for the few that dared to pursue Lamarckian experiments.

During the 1890s and the 1900s Ivan Pavlov, a highly regarded Russian scientist, conducted experiments on the conditioning responses of dogs. He used things like a bell or a tuning fork to alert dogs prior to feeding. After a time, the dogs started to salivate at the sound of the bell, even in the absence of food. Pavlov reasoned that the dogs had learnt to 'associate' the sound with the food as a reward, and that the new memories established were somehow inherited. Even though Pavlov claimed that it was easier to elicit a conditioned response with each successive generation, his experimental findings were negative.[10] Yet he continued to maintain his position on the inheritance of acquired characteristics until his death in 1936.

Although reports of Pavlov's experiments were not made available to western scientists until 1927, they became incredibly influential in the field of psychiatry by introducing the concept of 'associative learning'. Guide dogs, hunting dogs, sniffer dogs, retrievers, and guard dogs are all bred for different behaviours, based on the principles that Pavlov used in his early experiments.

In the 1920s, Austrian Paul Kammerer conducted experiments in an attempt to prove that Lamarck was right about inherited physical characteristics. Kammerer believed that he could take a variety of 'midwife toads' that normally mate on dry land and cause genetic changes in their offspring by raising them in water. The idea was that successive generations would develop so-called nuptial pads, typical of water-dwelling toads. The nuptial pads were needed for the midwife toad to successfully hold on to their mate in a wet and slippery environment.

Kammerer reported that nuptial pads did indeed appear on a few of the males. However, his political views were out of favour with the Nazis, and his work was discredited. When the test specimens were later examined, it was discovered that they had been altered using an ink dye. Although Kammerer protested his innocence, he committed suicide soon after.[11]

Another Lamarckian study that received a lot of attention was reported by William McDougall using rats. The first report of McDougall's Lamarckian experiments appeared in the prestigious British Journal of Psychology, in 1927.[12] Initially McDougall was interested in repeating Pavlov's experiments, using white rats in a small, but well equipped animal room at Harvard University. McDougall wrote to Pavlov and was surprised to receive a rather odd reply from him stating that he 'no longer held his deductions from his experiments to be valid'. However, McDougall continued with his research plan undeterred.

Putting the rats through their paces, McDougall reported that the first generation of rats made an average one hundred and sixty five mistakes on their first run through a maze.[13] After some practice they seemed to learn the route, and after a time were able to run through it perfectly. Once they learnt to do this, McDougall bred them and tested their offspring. This next generation of rats made an average of only twenty mistakes on their first run. He wrote that it appeared to him that Lamarckian transmission is 'a real process in nature'.

While the results of experiments reported by Pavlov and McDougall provided indirect evidence for the inheritance of newly acquired characteristics, neither scientist provided an explanation for the results they reported.

A decade later, the USSR's Trofim Lysenko attempted to prove Lamarckian inheritance. In the 1930s, Lysenko promoted the idea that crops could inherit acquired characteristics using a process called vernalisation. Vernalisation was the process of sowing winter wheat in the spring snow. Lysenko argued that the process of germinating wheat in the snow before planting would lead to greater crop yields. At the time, there was widespread starvation across Russia under militant communism, and the need for improved wheat production was acute. Lysenko had full state backing for his work and numerous scientists who opposed his findings were executed. Even so, the results promised by Lysenko were never delivered and no-one else was able to replicate the findings he claimed to have made. Despite this, Lysenko became an unassailable figure in the Soviet regime. In 1948 he became president of the Soviet Academy of Agricultural Sciences, and he retained his extraordinary intellectual authority until President Khrushchev's dismissal in 1964.

During the post-war period, up until Khrushchev's dismissal, there was an assumed close alliance between your political views and your scientific views: Western biologists who were members of a communist party were expected to support Lysenko. French geneticist, Marcel Prenant was such a party member. He initially attempted to steer clear of Lamarck and Lysenko, and was condemned by communists, non-communists and political commentators alike. In 1949 he felt obliged to state that only 'proletarian science' (i.e. that which supported Lysenko's work) could be right. Even that did not prevent him being expelled from the Communist Party in France.

During the post-war period, and up until the present, there has continued to be an assumed close alliance between your political views and your scientific views. The result was that any western scientist who crossed the boundaries between neo-Darwinian and Lamarckian thought quickly became the 'man left out'. It was during these years that neo-Darwinism truly reached its zenith. It remains a core part of our evolutionary belief system.

Only very, very recently has it become possible to put the neo-Darwinian period in context as just another episode in the overall history of evolutionary biology. The fact that we now know of several molecular pathways with distinctly Lamarckian implications means that we can begin to chart a new course for our evolutionary learning. At the very least, the neo-Darwinian view of the world needs to be updated to allow for the Lamarckian acquired inheritance phenomena that are beyond doubt.

It's now time to look again at our understanding of acquired inheritance effects in the context of the new genetics.

* * *

2. SOME HERITABLE GENETIC TRANSFER EFFECTS

When any organism incorporates new genetic material from another organism, the process is referred to as 'horizontal gene transfer'. Over half a century ago, when it was first suggested that horizontal gene transfer occurs, the idea was met with disbelief. One of the main reasons for not accepting the idea was that it implied that our genetic material is modular. That is, it might be flexible enough to enable new genetic information from another organism to be pasted into our genetic material. We use a similar process when we copy and paste text from one document to another in a word processor. However, with the advent of recombinant DNA technologies and the discovery of how bacteria swap genes, the idea has gained wide acceptance.

We now know that much of the DNA in almost all species—including humans—came from some form of genetic transfer effect. By comparing genetic sequence data, we have discovered that horizontal gene transfer involving nuclear DNA occurs regularly between complex organisms like humans. We also believe that it has occurred across all species many times in our evolutionary past. Clearly, our DNA has evolved as a living form of modular code that is able to accept new modules of genetic information from other organisms of the same or a different species.

From an evolutionary perspective, it seems that heritable genetic transfer effects are more significant than we previously thought, and that there are many ways that this can occur. In addition, we now have experimental evidence to suggest that when an organism incorporates new genetic information from another organism, offspring can sometimes inherit the new genetic cargo. When this occurs there are some obvious Lamarckian implications. It has also helped scientists to understand the mystery of how the speed and complexity of evolution appears to increase with time.

One of the more intriguing examples of a heritable genetic transfer effect is called the 'sire effect'. The idea is encapsulated in the following quote:

> *"My son looks like my wife's ex-husband."*

At first glance, the quote suggests sexual impropriety. It actually indicates something far more interesting.

MY SON LOOKS LIKE MY WIFE'S EX-HUSBAND

The sire effect is one of the more curious and the oldest forms of heritable genetic transfer effects described in the literature.[14] It is the belief that at least some of the genetic character of the sire that first breeds with a female can be passed on to her future offspring to another male. That is, a female's first sexual partner may serve as the genetic father of her offspring she bears in the future to another male.

This idea is ancient, and quite widespread. Animal breeders refer to this as 'throwing back'. But why has such a belief been held for so long, and among so many people?

Among the Chinese, stories of the son who resembles his mother's former husband have been circulating for hundreds of years. The effect has also been claimed in Italian folklore with the idea that the children of adulterous women are likely to resemble her husband, by being genetically influenced by the husband and her lover. Hence the Italian proverb offering women freedom in marriage: *filium ex adultera excusare matrem a culpa.* A translation is, 'a son born to an adulterous woman absolves her from blame'. It implies forgiveness by the son of an ex-adulterous woman.

Since the early 1700s, the sire effect has been used to guide the practice of English and European animal breeders.[15] Dog and sheep breeders still adhere to the idea, with some associations refusing to register stock whose mother was ever impregnated with a male of another breed, no matter how long ago. The effect has been used to improve some breeds of animal. For example, to breed dogs with more powerful jaws, breeders of Bedlington terriers mate a young bitch to a bull terrier and destroy the first litter of pups. They then mate the bitch to another pure Bedlington terrier in the belief that they are breeding an almost pure litter of Bedlington terrier pups with stronger jaws.

Many contemporaries of Charles Darwin wrote about the sire effect. In 1851, Dr Robert Balfour of Surinam was continually noticing among the native population that if a black skinned woman had a child to a white man, and then had children to a black man, the children generally had a fairer complexion than the parents. In the United States, the English social philosopher Herbert Spencer received a letter from a correspondent that said, "... children by white parents have been repeatedly observed to show traces of black blood when the woman had a previous connection with a negro."[16] Another nineteenth century observation on the sire effect is attributed to Australian scientist Count Strzelecki.[17] He wrote that in some indigenous populations of Australia and New Zealand, if a pure blood aborigine had a child by a European man, she lost the power to conceive a child by a man of her own race, but she could continue to produce children with a European partner. While such records are interesting to review in an historical context, it should be noted that the observations described are not scientific.

Putting aside the eugenic motives of some of these commentators, the idea that the sire effect occurred in humans was widely accepted among nineteenth century scientists. It gave rise to the acceptance of many of the eugenic practices later adopted by some governments in the twentieth century.

In keeping with the views of the day, Charles Darwin described many sire effects when he discussed the evolution of domestic animals. Darwin saw these as being of special importance in understanding evolution, and he used them as an integral part of his own theory of 'Pangenesis'.[18]

In one notorious example, he described the case of Lord Morton's Arabian chestnut mare which had her first foal to a quagga.[19] The quagga was a horse breed with zebra-like stripes that is now extinct. Her first offspring showed some of the markings and colour of the quagga. Later, wrote Darwin, when she was mated to a black Arabian horse, she produced two colts that had more distinct stripes on their legs than the original offspring from the quagga. One of the colts had stripes on his neck and some other body parts, and his mane was short and stiff like that of a quagga. Darwin also described the example of Lord Western's sow that first mated with a chestnut coloured wild boar. Afterwards, when the sow was mated with her own breed, which was known to be a very true breed at the time, some of the young pigs were marked with the chestnut colour of the wild boar.[20]

In later evolutionary debates, the sire effect was used as a major weapon against the idea of Weismann's barrier, but it gradually lost respect among twentieth century biologists. Yet, even Weismann felt compelled to put forward an explanation of the sire effect without violating his own ideas on germplasm.

While Weismann's barrier was conceived to prevent any exchange of genetic material between the body (or soma), and the reproductive sex cells, he also described the possibility of sperm directly penetrating the immature ovaries. This meant that the genetic material in sperm could not only penetrate the ova used in immediate fertilization, but that the genetic material of some sperm could also penetrate immature ova. When mature, these ova could then accept further genetic material from a new partner at conception. In putting forward his idea of genetic exchange, Weismann described the first cellular model that explained how offspring could inherit a mix of genetic material from the male partner of the mother at conception, and from her previous partners. He was satisfied that this possibility did not violate his idea that the contents of the reproductive cells do not change from one generation to the next.

In the early 1980s, the phenomenon was observed in the immune system. In experiments by Reg Gorczynski and his colleagues in Canada, they used a group of normal female mice that had at least three litters with males that had received a skin graft and been inoculated so that graft rejection was significantly delayed. They showed that these females continued to produce some offspring showing the same delayed rejection of a skin graft, even when the new father was a normal non-grafted male.[21] In these experiments, the male sperm somehow immunized the mothers and their subsequent offspring. The results were dismissed with some scepticism by geneticists because of the continued belief in Weismann's barrier.

In 1986 further experiments verifying sire effect phenomena were conducted using rabbits. As the wild rabbit populations increased they caused damage to farming lands in countries like Australia and England. To control the rabbit population in Australia in the 1950s, scientists working at the Commonwealth Scientific and Industrial Research Organisation (CSIRO) released rabbits infected with the *Myxomatosis* virus into the wild rabbit population. While the virus was effective at controlling the wild rabbit population initially, scientists became concerned about the speed with which the population was able to develop a natural resistance to the virus.

To understand how the rapid resistance was generated within a wild population, Bill Sobey and Dorothy Connolly conducted a number of controlled breeding experiments at the CSIRO.[22] They reported evidence which suggested that bucks which had acquired immunity were able to pass on their immunity to offspring when mated to a doe who had not previously been exposed to the *Myxomatosis* virus. That is, Sobey and Connolly found an unexpected and heritable paternal transmission effect to their offspring. The immunity the bucks had acquired was passed onto a large proportion of offspring. When they mated a non-immune buck to a doe that had previously been mated to an immune buck, they also found that some of the offspring were born with immunity to the virus. While these results helped to explain the rapid increase in immunity observed in wild rabbit populations, Sobey and Connolly could not explain the data using conventional genetic explanations. They concluded that an unknown factor appeared to be transmitted via the semen of the recovered bucks to the does, and that the unknown factor was nongenetic in nature.

The scientific evidence reported by Gorczynski and his colleagues, and Sobey and Connolly are important in that they provide scientific support for two very different but related phenomena: Lamarckism and the sire effect. Both challenge the neo-Darwinian orthodoxy.

Sire effect phenomena have been squarely ignored by scientists since these experiments. This is despite the fact that it is now possible to evaluate any changes using genetic markers, rather than relying on observations of differences in physical and biochemical features, as the initial experiments required. Because there have been no further experiments, we still do not understand how the effect occurs. Yet we know that sperm cells do penetrate into the soft surrounding tissues of the female genital tract. We know that some of the DNA (the nuclear storehouse of genetic information) and the RNA (the genetic guardian processing how DNA is used) are both released by sperm cells and can become integrated into the female somatic cells. During sex, millions of sperm cells containing DNA, RNA and proteins are deposited into the body of a female, and much of it is absorbed by her.[23] Sperm have been found to penetrate into the uterus wall and glands, the cervix, the fallopian tubes and the ovary walls. It has been found that the cells of younger females are more permeable to foreign sperm DNA than the cells of older females. There is also evidence to suggest that the lymphatic system is used to carry sperm to other maternal tissue cells, including the heart and brain. Further investigation may find that the DNA and RNA cargo associated with sperm cells also penetrates male somatic cells during male to female sex, male to male sex, or during oral sex. On current evidence, there is a low take-up of this foreign DNA by somatic cells, although much more experimental work is required here too.

We also know that foetal DNA is present in maternal blood, and that this provides a possible alternate route for sperm DNA and RNA to become embedded into the female somatic cells. The seminal fluid of fertile men is known to contain thousands of different RNAs as well as DNA. We do not yet know the role of most of these. We know that the amount of foetal DNA and RNA present in maternal blood increases as the pregnancy progresses. The foetal DNA circulating in the mother's blood is now used to provide a non-invasive way for genetically testing a foetus before it is born. This makes it possible to genetically screen a foetus for a wide range of ailments without the risk of abortion. The sex of a foetus can be identified as early as six weeks into pregnancy. Genetic screening for a wide range of genetic defects such as Down's Syndrome or sickle-cell anaemia can also be conducted using a small sample of maternal blood.

All of these represent viable natural pathways for the transmission of sire effects to the next generation.

While the technologies are available to conduct further genetic research in the area, maybe we are not ready to understand the full impact of such effects? The sire effect has some very deep social implications related to teenage sex, rape and remarriage. How many men would be happy to have children that look like their wife's previous lover? How many women would be happy to learn that in absorbing the genetic components of the sperm of their husband, their genetic profile could be altered to become more like that of their husband's over time?

A lot more direct experimental work is required to understand the molecular mechanisms at play, the regulatory processes involved, and the extent or circumstances that allow the effects to occur (or not). Until then, we can't even begin to address some of the wider social, political and evolutionary implications.

PEKIN DUCKLINGS

Some of the best known heritable genetic transfer experiments resulting in morphogenetic changes involve ducks. One of the first experiments using ducks and showing that foreign DNA could be introduced and then integrated permanently into a new vertebrate host was reported in 1960.[24]

The story began when French biologists Jacques Benoit and Pierre Leroy took twelve purebred Pekin ducklings and injected them with DNA that was extracted from the genitals of Khaki Campbell ducks. They were amazed to find that when the ducklings matured, the majority of their offspring developed a range of characters derived from the Khaki Campbell ducks. Many of the offspring showed a combination of typical Khaki Campbell features that included beak pigmentation, feather type, head shape, and body shape. They had the Khaki Campbells' greenish coloured bills, and their feathers were quite unlike that of their parents.

When Benoit and Leroy saw the first generation of strange looking ducklings, one of the first questions they asked was: Is the change permanent? To answer this question they conducted further breeding experiments. They found that the injected Pekin ducks continued to produce the strange ducklings. They also discovered that eighteen second generation ducklings out of twenty six showed Khaki Campbell duck features. They concluded that they had basically altered the parents' genetic characteristics.

When the first results were announced to the world, many were stunned. In Time Magazine the renowned French biologist Jean Rostand drew attention to the disturbing possible implications of 'transforming one race to another'.[25]

Although the results proved difficult to repeat in a range of other animals, including rats and rabbits, the same effects were found in *Ephestia,* a species of small moth. Pure DNA from wild-type animals was injected into recessive larvae. When they matured, they developed wings with the same coloration as that of the wild-type animals.[26] The features of the wild-type appeared in animals in the treated generation, and in subsequent back-cross generations.

We still do not fully understand why some of the early experiments failed, while others succeeded. Yet the experiments were important: they demonstrated for the first time that DNA transferred from one species of vertebrate to another could become integrated into the genetic material of body cells and the germline—and then passed on to the next generation.

A PINK IRIDESCENT GLO-CHIHUAHU

By the early 1980s genetic tools enabling foreign DNA to be integrated into the DNA of an organism had been developed.[27] This marked the start of the recombinant DNA revolution. The new recombinant genetic tools and techniques enabled scientists to begin to efficiently transfer foreign genetic material into the DNA of animals. Only then was it possible for major advances to be made in understanding the nature of some of the heritable genetic transfer experiments described so far.

In the process of conducting the early trials, scientists are finding that the genome of animals can readily integrate new genes, or have selected genes deleted. To their surprise, they are also finding that if the new genetic material is transferred to somatic cells, the progeny of new transgenic animals inherit the new genetic information.

Patrick Fogarty, of the biopharmaceutical company Tosk Inc. in California, is one of a number of scientists working with animals to develop new genetic transfer technologies for drug target discovery. In his experiments, transgenic animals that have 'knock-in' genes or 'knock-out' genes are used to provide useful animal models for the development of gene therapies.

Fogarty and his colleagues developed two new delivery vectors that are used to incorporate a new gene into a genome, or to replace a similar gene with a new one. In the laboratory he uses a mechanism to allow

him to inject foreign DNA into the tails of mice. Each
animal is directly injected with a mix containing the
new gene and a vector designed to assist the integration
of that gene into the animal's own genetic makeup. The
targeted gene delivery (TGD) and the StealthGene™
vectors used by Fogarty are a modified piece of DNA
that is designed to specifically target, then insert or
delete additional genetic information into the DNA. We
have known for a long time that all animals and plants
have transposable segments of DNA that can move from
one location to another. Yet it has only become possible
to use these to deliver new genetic information into the
DNA in the last two decades.

In his experiments, Fogarty injected the vectors with
the new DNA cargo into mice. He showed that over
time, the new DNA became integrated into almost all of
the somatic cells tested. There also appeared to be no
side effects using this technique.

What is most remarkable about this work, is that
the offspring of the TGD and StealthGene™ transgenic
animals were also transgenic.[28] The new genetic
material had not only entered the somatic cells, it
had also altered the genome of the animals injected.
Using only the male line, the new genetic information
was inherited by twenty five to eighty percent of the
progeny of the transgenic animals, depending on the
test variables used. The new genome was stable for
the first four generations tested, and there appeared to
be no strain or sex dependencies.

Another unexpected finding was the speed with
which the genome of the mice integrated the new genetic
material. Fogarty and his colleagues examined progeny
from animals receiving the TGD or the StealthGene™

vectors that were mated as early as two days, and up to one year, after the injection was administered. Quite unexpectedly, transgenic progeny were found in all groups. While the reason for the fast uptake is not yet fully understood, Fogarty hypothesised that both the mature germ cells existing at the time of the injection and the germline stem cells readily integrated the new genetic cargo.[29]

So it's clear that foreign DNA can become rapidly integrated into the somatic DNA and the germline DNA in mature sperm or ova. But we do not yet understand all of the regulatory mechanisms involved. This is the fascinating problem of the Gatekeeper: what regulates what new genetic information gets in? Clearly nature has a lot of tricks to prevent our genome from being completely open to foreign updates, and it has invested in some sophisticated regulatory mechanisms to limit the type and amount of new genetic information that can be integrated. However, once the genome sequence is altered, the changes can be passed on to future generations.

The work of Fogarty and many others has now set the stage for the use of genetic engineering technologies to conceive of and produce a vast new range of genetically modified life forms. The implications are profound. Humanity has found new powers to not only exert a greater control over nature, but to become the creator of new life forms. With the ability to transfer genes readily across species in the laboratory, we are entering into a world where we can achieve the wholesale horizontal transfer of heritable genetic information at will. We also have the technologies to rapidly design and build new genetic sequences. However, it is likely that experiments involving the insertion of genetic material directly into human sperm and ova will remain off-limits by law for most of the world's scientists. At least for the present.

Yet people are already devoting their lives to breeding transgenic fish, pigeons, dogs, or roses. We do not have to go further than our supermarket to see how rapidly genetic engineering is altering our natural world. We can buy bread made using a genetically engineered grain.

There are already many industries that rely on the use of genetic engineering. We have the technologies to rapidly synthesize and clone vast quantities of a target DNA sequence. Work has already begun on the development of table-top sequencers able to clone a whole genome in a few hours or less—instead of the thirteen years it took to sequence the human genome just a few years ago. The same technologies might be used to produce personalised gene therapies to cure or prevent common cancers, or to improve our own genetic profile before we have children.

This raises some questions about the more general use of new genetic engineering tools as many of the techniques used can now be fully automated. How long will it be before you can order a pink iridescent glochihuahu from your local 'Creative Pets' store?

Theoretical physicist and futurist Freeman Dyson refers to these developments as 'the domestication of biotechnology'.[30] Dyson sees a future where people will be able to acquire a do-it-yourself portable gene synthesizer that will enable them to produce a new DNA sequence. Using a DNA template downloaded from the web, and a home gene kit, they would be able to set about breeding a new designer pet.

Dyson argues that the tools of genetic engineering will remain unpopular while they are in the hands of companies like Monsanto, who are developing and patenting new genetic crop strains. He compares the future domestication of genetic engineering technologies to the process whereby computers were introduced into millions of homes around the world, and the impact this has had on society. He likens the current environment to the early days of computing when a handful of experts controlled the world's entire computing power.

GENE THERAPY

Gene transfer technologies are rapidly being adapted to provide us with new ways to treat and manage human diseases. Already gene transfer technologies are used to create a range of gene-based vaccines.[31] To avoid Orwellian connotations associated with eugenics the terms 'gene therapy' or 'genetic therapy' are used here to describe the use of new gene transfer technologies in medicine.

Gene transfer technologies also provide new more compatible sources of human tissue for transplants. We could grow our own new cells to replace dead or damaged cells due to cardiac arrest, brain damage, spinal damage, abnormal brain changes (such as in Alzheimer's patients) or damaged tendons. We might even be able to improve the quality of our own lives, and extend our lifespan by another decade or two in the near future. The possibility of developing a genetic maintenance program for each of us could help us to avert certain diseases. Our genome could conceivably be updated so that we can function optimally for longer.

We can already sequence, synthesise, cut and paste new genetic sequences into vaccines to combat several forms of cancer, including cervical cancer, renal cancer and melanoma. Still, "Why can't they develop a cure for cancer?" remains a very common question. The simple answer is that cancer is actually a wide range of diseases. The development of genetically engineered cancer vaccines is still in its infancy, and the jury is still out on their effectiveness and safety. However, by understanding the genetic profile of an individual and any forms of cancer they might have, we will be able to produce personalised genetic inoculations made up of tumour cells that provide the immune system with an alert to attack against any related cancers. While we are still a few years away from being able to do this for humans, the research is promising. Meanwhile, many other genetic approaches are being investigated. All of these early efforts rely on the ability of our genetic makeup to integrate the new DNA, and then to reproduce the new sequences in all new cells our bodies produce.

Gene transfer technology also has the potential to be used in other more subtle aspects of our lives such as to modify our behaviour, or reduce our susceptibility to certain diseases, and to forecast a lot more about our health than any of us might be able to guess at present. It could also be used more widely to alter the genetic profile of a foetus, or a newborn baby. The genetically engineered medicines and diagnostic technologies are likely to be safer and less invasive than many current treatments. They will be more targeted and designed to match the specific genetic profile of each of us.

Yet, there remains an uneasiness among many as we enter a world that we still know very little about. After all, we once thought that dumping chemicals like DDT into our environment could save the world from starvation. It seemed like a great idea at the time.

Gene transfer among humans also occurs quite naturally among us. The realization of the number of ways horizontal RNA or DNA transfer can result in infectious diseases has raised concerns among biomedical scientists. Every time we kiss, shake hands, or have unprotected sex, humans are in fact conjugating like bacteria. For example, we can acquire a new disease such as hepatitis C, glandular fever, a new strain of herpes or just a plain old cold as we absorb viral material from others. As we interact with others, we transfer a lot of new viruses that could become integrated into our own DNA. When we exchange bodily fluids, a lot of genetic material is freely exchanged. There is a very high concentration of RNA viruses in semen, and to a lesser degree in saliva. Some of these could be harmful slow acting RNA viruses (*retroviruses*) that may take decades to cause symptoms. We still do not fully understand the genetic and evolutionary consequences of such large scale horizontal transfer effects among humans. Horizontal gene transfer from human to human, and the possibility of vertically transferring new genetic material to the next generation is an area that is largely ignored in the literature.

The effect of extensive transfer of genomic material also complicates the ability of scientists to conclusively map evolutionary trees based on the sequence of one gene or genetic markers. Based on morphological differences, we once believed that the ascent of all species through time was expressible as a metaphorical tree of life with branch tips representing current living forms, and all leading to a common ancestor. Using genetics, scientists have tried to trace our ancestry back to a primal female ancestor—the African Eve. The view now emerging is that the tree of life gives us an incomplete picture of our genetic evolution. It cannot explain the full complexity introduced by large scale gene transfer effects. In fact, the notion of a 'genomic tree' no longer fits data arising from recent research. The genomic tree is far too intricate in detail to be able to simply plot gene transfer events from one species to another, or from one generation to another.

In summary, few now doubt that new genetic engineering technologies are giving rise to one of the biggest biomedical revolutions so far. The idea that we can cut and paste genetic material into our somatic and germline cells at will represents a huge untapped potential for the decades ahead. The finding that at least some of the new genetic material is passed on to the next generation will have a huge impact on our future evolution: it will have a bigger impact on our lives and our longevity than the introduction of penicillin. However, while the future of new genetic transfer technologies will be big in terms of its size, and the implications for medicine and our evolution, we do not know if it will result in a blurring of the boundaries between species, nor the pathways involved in altering the genetic makeup of future generations. In the meantime, a lot more research on the acquired inheritance effects is required. Some of the research has focused on the microbial world.

GENE TRANSFER AND ADAPTIVE MUTATION IN MICROBES

It is well known that horizontal gene transfer effects occur frequently between bacteria, between fungi, and between bacteria and some fungi, and that these play a key role in shaping their evolution.

In 1944 Oswald MacLeod and Maclyn McCarty stunned the scientific world. They reported the world's first significant example of a heritable and reproducible genetic transfer experiment. They showed that when you feed a non-encapsulated pneumonococcus bacterial strain with pure nucleic acid from a more virulent and lethal encapsulated strain that has a protective coating on its surface, the bacteria that were fed the nucleic acid transformed into a more lethal fully encapsulated strain.[32] These experiments demonstrated that it is the nucleic acids that determine the bacterial protein structure. MacLeod and McCarty also showed that changes induced in the bacteria are predictable and heritable.

To conduct their experiments they extracted around ninety litres of nucleic acid material from pneumococcal bacteria, which they called the 'transforming factor'. They went to extraordinary lengths to show that it was purified nucleic acid, and not protein or carbohydrate. At the time, it was believed that protein was responsible for hereditary characteristics. Most scientists did not believe that it was possible to introduce and inherit new genetic material using this approach.

This was groundbreaking research as it showed that not only were the changes due to the DNA material, but that it provided the first evidence of the importance of genetic transfer between organisms.

By the late 1950s, horizontal gene transfer effects among bacteria of the same or different species were well known: the genes that conferred antibiotic resistance on one type of bacteria could be horizontally transferred to another species of bacteria. The second species became genetically altered so that it too showed antibiotic resistance.

Bacteria have evolved several mechanisms to protect themselves against antibiotics. The most common method is to enzymatically alter the antibiotic so that it becomes ineffective. Another strategy is to physically remove it from the cell. When one bacterial cell in a human acquires antibiotic resistance, it can quickly transfer that resistance to many other different types of bacteria—even those that are only distantly related. The medical significance of discovering that bacteria can quickly develop resistance to antibiotics was immediately recognised by scientists: they also recognized that the war between our immune system and antibiotic resistant bacteria had begun!

The rate at which bacteria become resistant to antibiotics is being accelerated by the almost worldwide practice of adding antibiotics to animal feed, and the overuse of antibiotics by doctors. The effect of eating meat from animals raised on food with antibiotics in it is comparable to being on a mild course of antibiotics all of our lives. This creates an environment where it is almost inevitable that more bacterial strains of diseases such as tuberculosis could become incurable—just as they were before antibiotics were developed.

Investigations into the evolution of antibiotic resistance in bacteria exposed to low concentrations of antibiotic have shown that it is the result of a combination of horizontal gene transfer effects, some DNA mutations, as well as a number of different molecular and genetic inheritance phenomena as the bacteria adapt to their new environment.[33] Together, these provide the bacteria with a powerful arsenal of molecular machinery that enables them to outsmart whatever new drugs we are able to develop to combat them.

Since the 1942 introduction of penicillin to fight bacterial infections, a broad range of new antibiotics with a number of different chemical structures have been developed. However, the accelerated evolution of bacteria exposed to these drugs means there is bacterial resistance to every approved antibiotic. Some bacteria have emerged that are even resistant to the most powerful multi-drug barrage approaches. By fighting them harder, we are actually sending their evolutionary molecular machinery into overdrive. The result is the persistent and rapid development of drug-resistant strains of dangerous bacteria.

To overcome our shrinking arsenal of effective antibiotics, scientists are focusing on developing new strategies. However, to be successful we need to know more about how the bacteria evolve resistance so rapidly. The hope is that we can use alternate drugs to interfere with these processes at key points in the molecular pathways. One approach is to chemically alter the bacteria so that our own immune system is able to clear the infection. Other approaches involve altering the bacteria themselves so that they are no longer resistant to traditional antibiotics. Some researchers are looking at ways to prevent the horizontal DNA transfer that occurs between bacteria. However, these approaches have only just begun, and a lot more research is required.

Meanwhile, the rapid evolution of bacteria has sparked debate about whether the process of adaptation is due to the forces of natural selection alone, or acquired inheritance effects driven by environmental stimuli. Some scientists have argued that acquired inheritance can be observed among single-celled organisms, but this view is still not widely accepted.

As far back as 1988, John Cairns and a group of other scientists at Oxford University published a paper in the prestigious journal Nature, in which they concluded that they had found evidence suggesting that bacteria could somehow select which mutations to produce.[34] In the experiments conducted by John Cairns and his group, they took a sample of the bacteria *E. coli* that was unable to consume lactose, and placed it in an environment where lactose was the only food source available. They observed that the genetic makeup of the bacteria rapidly changed so that the original bacteria and the next generation of bacteria were able to use lactose. The bacteria use various forms of hypermutation plus trial and error to enhance their search for the most appropriate mutations. However, no one is claiming that bacteria can intelligently change the specific base needed to respond to a specific environmental challenge analogous to the way a human genetic engineer might intelligently make just those precise changes needed.

If the evidence is beyond doubt, why is there such a strong resistance to the idea of acquired inheritance effects in bacteria among so many scientists? Unfortunately, it seems to come back to ideology, rather than evidence. At the time of Cairn's experiments, Lamarckian views were still frowned upon by the scientific establishment. To avoid any unwanted Lamarckian implications, Cairns referred to the processes described as 'adaptive mutagenesis' so that the work could be viewed from an ontogenic perspective. In other words, he wanted the results to be viewed from the perspective of changes that occur in an individual cell rather than a population that changes over a much longer period of time (*phylogeny*).

Another group of scientists working with yeast has discovered that yeast cells containing a specific protein, Sup35, are able to take up new genetic material, some of which gave them new abilities such as resistance to a particular herbicide. When the yeast cells with the protein Sup35 were bred with genetically identical cells not containing the protein, the trait reappeared in some of the resulting offspring.[35] The scientists had demonstrated that a protein-based form of acquired inheritance occurs in yeast without altering the DNA sequence of the genome.

Experiments using other test systems and organisms have since confirmed that protein-based acquired inheritance phenomena are widespread. They occur in single celled organisms like yeast and bacteria, and also in plants and mammals. The idea that some special types of protein (called 'prions') can transmit heritable information directly from one protein molecule to another is relatively new. This is a form of autocatalysis— one prion catalyses the formation of others. While this effect is not strictly a gene transfer effect, the wider implications for our ideas on evolution, genetics and medicine are profound: the findings are causing us to rethink the nature of what 'heritable material' is and how heritable information is shared between species, between individuals of the same species and subsequent generations. However, much further research is required to fully understand the implications by studying the molecular pathways involved.

Thus, by studying bacteria, and other single-celled organisms in the microbial world, we are starting to find some quite unexpected mechanisms that alter our view of how acquired inheritance phenomena might occur. In cases involving genetic change, microbes seem to have the ability to produce genetic variations at times of environmental stress, and in gene locations that are most likely to be involved in reducing the effect of the stressful situation (from the microbe's point of view). It is absolutely established that the bacterium can mutate just the genes it needs to recover a function. While a bacterium cannot possibly know that it needs to mutate a specific gene, it does have the molecular machinery to respond to environmental stress signals in a way that triggers a range of genetic transformations. Some of the mechanisms involved are described in the Chapters following.

In cases involving molecular changes that alter how the same genes are expressed, the microbes also have the ability to transfer and inherit non-genetic information. As a consequence, the cellular and molecular genetics literature is now sprinkled with references to the 'Lamarckian'[36] nature of the phenomena described.

GM CROPS, SUPER WEEDS AND SUPER BUGS

> *"You may drive out Nature with a pitchfork, yet she will ever hurry back, and, ere you know it, will burst through your foolish contempt in triumph".*
> Horace, Epistle I-10s.

Lamarckian inheritance phenomena have been known to occur in plants for many years. The concept of Weismann's Barrier has never been extended to plants because somatic modifications in plants are propagated in the seeds which are formed by the parts of the plant responsible for generating any genetic changes. We can all watch the gradual development of seeds in flowers like sunflowers, or the formation of almonds on an almond tree.

One of the pioneers of genetic inheritance effects in plants was Barbara McClintock. She began her research using maize in the 1940s, and later discovered what she called 'jumping genes'.[37] Jumping genes enhance the search for adaptive responses to stressful environmental conditions. In recognition of the significance of this work, she was awarded the Nobel Prize in Physiology or Medicine. Her experiments showed that plant genes had the ability to rapidly adapt, and that the seeds that produced the next generation of plants carried the same genetic changes.

Today we use 'gene gun' technologies to transfer new genetic information into plants. This is faster, and it results in a highly targeted genetic change. Gene gun technologies use tiny gold pellets that are coated with foreign DNA cargo. The pellets are fired into the chromosomes of a plant they wish to genetically alter. The plant cells have the genetic and molecular machinery required to integrate the new DNA into their chromosomes. After careful cultivation, the next generations of plants with the new genes are produced.

Gene gun technologies are now routinely used to produce improved strains of corn, sunflower seeds, tomatoes and cotton that have their own built-in insecticide. They can also be used to make crops like tomatoes that retain a fresh look for longer on a supermarket shelf. Scientists are also working at developing coffee beans that do not contain caffeine, potatoes and rice with more starch, or beans that contain more protein.

While some of the commercial benefits for agriculture and transportation of food over long distance are obvious, there are also some good reasons for caution among consumers. The possibility of breeding superweeds is just one such justified concern.

Allison Snow, a Professor of Evolution, Ecology and Organisational Biology at the Ohio State University was the first to show that genes artificially inserted into crop plants can migrate to weeds nearby by a natural process of horizontal gene transfer that occurs among plants. Her work showed that artificially made genetic information that was inserted into one plant species to fend off insects could be horizontally transferred to weeds in a natural environment. Weeds in a natural environment that also fend off insects could potentially become a greater environmental problem. What chemicals or other genetic tools would we then need to develop to control a new generation of highly vigorous weeds?

In one study, Alison Snow and her colleagues used sunflowers that had been genetically modified to produce a chemical that is toxic to certain insects. The resulting hybrid sunflowers produced around fifty percent more seeds than the unmodified control plants. The implanted genes also became integrated into the weeds growing nearby. It was already known that genetically modified crops could potentially cross breed with nearby weeds to produce 'super weeds'. Demonstration of the transfer of new transgenes to the wild plant population has resulted in such experiments being halted—for now.

In looking at the possibility of growing superweeds with an inbuilt insecticide, we have to take into account that nature also has the tools to rapidly produce a new generation of 'super bugs' that can tolerate the insecticides in the super weeds. Once the 'super bugs' are produced, what new chemicals will we need to introduce as insecticides, and what other farming practices will we need to use to ensure that plagues of superbugs don't also eat other crops or enter our gardens, and cause further destruction?

These technologies provide a new set of powerful tools, but their use is strongly questioned by many. The Director of the Institute of Science in Society (ISIS) in London, Mae-Wan Ho, is a pioneer of the physics of organisms and is one of the world's most influential writers warning against the dangers of genetic engineering.[38] As a leading advocate for Lamarckian inheritance effects, she provides some of our most significant insights into the potential hazards of growing genetically modified crops without fully understanding what we are releasing into the environment. Our inability to control horizontal gene transfer through cross-pollination between genetically modified crops and natural strains is just one of the dangers she highlights. She writes about cows that have died after eating genetically modified maize, and rats that are stunted, dead or sterile after being fed on genetically modified soy.

While it is right for the public and governments to be more fully informed about the dangers of genetic engineering, we also need to continue the debates based on sound scientific information. To do this, all governments need to nurture a research capability that is not corrupted by the influence of industry funds, and free to investigate and question all possibilities in relation to evolutionary thought. We need to invest far more for understanding heritable gene transfer effects that occur naturally in nature, and the forces that nature could unleash against us if we fail to do this.

* * *

No one has challenged the validity of any of the heritable genetic transfer phenomena—which have some distinctly Lamarckian implications. Heritable horizontal genetic transfer effects can and do occur in fungi, bacteria, plants, frogs, flies, ducks, mice and men. All living things have the genetic and molecular machinery to enable foreign DNA or RNA to become integrated into new somatic cells and the germline ready for the next generation to inherit. While we do not yet understand all of the possible pathways, we are developing the technologies that will enable us to investigate the underlying mechanisms in more detail.

From an evolutionary point of view, it seems likely that horizontal transfer effects have played a significant role in our evolutionary history. In all life forms, the ability to integrate foreign DNA has given rise to some of the extraordinary diversity we see in living forms. It seems that the genome of all living things has evolved to benefit from diversity partly due to horizontal gene transfer effects.

But there are also a number of other ways that acquired inheritance effects can be generated in our lifetime.

* * *

3. SPERM MEDIATED GENE TRANSFER EFFECTS

When Albrecht von Kolliker traced the genesis of sperm cells and proved that they arose as differentiated tissue cells in 1941, few realised the potential evolutionary implications of this work. During the intervening years, scientists continued to believe that the heritable information in the somatic cells and the reproductive germline cells does not mix, thus preserving the original integrity of our genes. We believed that the hereditary information contained in sperm cells remained the same throughout life. We also believed that in females, the ova are all present at birth, and that the hereditary information they contain is unchanged throughout life.

In the last decade we have learnt that this is not the case. We are discovering some additional sperm mediated gene transfer effects operating that could have a large impact on the next generation. It is now well known that there are stem cells in bone marrow that are used to differentiate into ova in adult females and sperm cells in adult males. We are continuously growing new germ cells throughout our adult life.[39] Somatically derived DNA and RNA enter the germline DNA and RNA when new sperm cells are being formed from the stem cells. When the new cells are being formed in the bone marrow, they are able to integrate somatically updated DNA, and so confer the benefits of new genetic information to the next generation.

We also know that as soon as they are formed, the new sperm cells can act as a vector for carrying additional foreign DNA and RNA fragments—like tiny passengers attached to its surface. Once attached, sperm cells can ferry the additional cargo directly to other sites, including the female genital tract and into an ovum at fertilization. It seems that the naked nuclear DNA of sperm is not the only one invited to the conception party!

Another important sperm mediated gene transfer effect we have discovered in recent years is that sperm cells are particularly vulnerable to exposure to chemicals in the weeks prior to fertilization. The impact can be swift, and the effects are inheritable.

GENETIC HITCHHIKERS ON THE SURFACE OF SPERM

Research by scientists including Corrado Spadafora, working at *Instituto Superiore di Sanita* in Rome, has done much to expand our knowledge of the molecular mechanisms involved in sperm mediated gene transfer. Spadafora and his team have established that sperm cells act as a vector for carrying foreign genetic sequences into the next generation. When this occurs, it provides an opportunity for foreign DNA or RNA (nucleic acid) cargo to be transmitted by the father to the offspring, with some important implications for evolution.[40] It is now well established that mature sperm cells in almost all animals have at least one mechanism for carrying foreign DNA or RNA into a new ovum at fertilization. One method used involves foreign DNA and RNA molecules attaching themselves to the sperm surface. Like genetic hitchhikers they get a free ride to where all the action happens during fertilization of an ovum, and some can become integrated into the sperm nuclear DNA prior to fertilization. Here's a simple description of how this works.

Seminal fluid is known to contain a lot of nucleic acids and proteins. It is loaded with retroviruses, some harmful and some not. Ejaculation fluid is literally a retroviral injection, with a virus titre greater than 10^{14} particles/ml. Only the placenta is known to have a similarly high concentration of retroviruses. As a result, a lot of additional RNA or DNA can become bound to the surface of sperm cells in seminal fluid.

The act of binding to the surface of sperm cells triggers a series of enzymatic functions. We don't yet know every step of the process, but some investigation has been done into the copying of the foreign DNA or RNA into the sperm cells. Once inside, a small proportion of the foreign sequences may become inserted into the DNA in the sperm at selected sites prior to fertilization.[41] When foreign RNA or RNA is copied into the DNA of the sperm cell, the chromosomal DNA of the sperm cell is rearranged to accommodate the new DNA. Integration can occur at one, or several specific locations of the chromosomal DNA structures in a sperm cell. Some of the newly inserted DNA sequence may then become inherited by progeny.

This suggests that the DNA and RNA inside a sperm can potentially be endlessly rewritten and rearranged as each new strand of DNA or RNA attaches itself to the surface. This is like watching an army rapidly rearrange their artillery when a new set of instructions is received by a General on the battlefield. While we know that at least some foreign DNA and RNA can become integrated into the sperm's DNA, we don't know how frequently, or under what circumstances, these transfers occur in nature. But just because we don't yet understand it, doesn't mean there isn't a pattern to it. The binding and copying of the foreign RNA or DNA are highly regulated processes, and carefully mediated by a number of molecular cofactors that are present within the sperm cell itself.

In laboratory experiments using mice, sperm are used as gene vectors to insert a foreign nucleic acid sequence into offspring and so create transgenic mice. Human sperm also have the same enzymatic activity as mice, although we still have not characterised all of the possible pathways that this mechanism might use, or the other molecular cofactors involved.

SMOKING CAN CAUSE SPERM DAMAGE

During some of the later stages of the process of maturation, male sperm cells become very sensitive to environmental chemicals that can cause mutations. This is because maturing sperm progressively lose their ability to repair any DNA damage. We do not know why. During this repair-deficient phase when the normal cellular repair processes are unable to repair any damage, the maturing sperm cells are very vulnerable: it is during this phase that immature sperm can accumulate significant heritable genomic damage.

When mice are given daily injections of chemicals from tobacco smoke at different stages prior to conception, we see just how vulnerable maturing sperm are to genetic damage due to smoking. In the experiments reported by Francesco Marchetti and Andrew Wryobek working in Berkeley in California, sperm accumulated heritable DNA damage. And most of the damage was accumulated in the few weeks just prior to maturity when fertilization would occur.[42] They also reported that a few of the fertilizing sperm had unusually high levels of chromosomal damage. What was surprising was that continuous exposure to low doses of the tobacco chemicals prior to conception was found to be just as detrimental as high level or acute exposure over the same period. So smoking is detrimental for males just prior to conception—regardless of how many cigarettes they smoke each day!

Comparative studies using ova found that the ova did not accumulate the same high levels of damage. The experimenters concluded that this was because the repair mechanisms in ova are fully functional during all stages of maturation.

These experiments and some similar ones have now established that environmental mutagens like industrial chemicals and cigarette smoke are all likely to cause the greatest damage to DNA in sperm during the last few weeks of maturation. This makes young males particularly vulnerable to habits like smoking and the effects of chemicals in the workplace in the weeks prior to conception.

HEALTH RISK AND EVOLUTIONARY CONSEQUENCES

Any mechanism that results in foreign DNA or RNA being taken up by sperm cells and then integrated at fertilization presents a high risk factor for human health. Foreign DNA or RNA from a male can easily be taken up and integrated by the sexual partner or a new embryo. It also means that partners and offspring become particularly vulnerable to sexually transmitted viruses such as HIV, human papiloma virus (HPV) or herpes.

Heritable sperm mediated effects also makes males particularly vulnerable to exposure to a variety of chemicals just prior to conception. Young males who are exposed to chemicals at work, who are smokers, who are recreational drug takers, and those on some types of medications might all be putting the next generation at risk—usually without even knowing. Given what's at stake, and how little we understand of the big picture, we need much more research in the area.

* * *

4. THE FLUID NATURE OF OUR HEREDITARY MATERIAL

The birth of modern genetics came in 1952 when A. Hershey and M. Chase confirmed that DNA was the hereditary material. Just one year later, James Watson and Francis Crick discovered that our DNA is nucleic acid composed of two chains of bases A (adenine), G (guanine), C (cytosine) and T (thymine). The two chains are joined together by a ladder of hydrogen bonds, and shaped into a double helix. In the complimentary strand, A always pairs with T, and G always pairs with C. These are the simple and universal pairing rules for the DNA of all living organisms.

Once it was established that all cells contain DNA in their nucleus, we assumed that DNA alone provides a set of universal building blocks for all known living forms—the only exception being bacteria, which has DNA but no nucleus. This is what children have been taught in class for decades: DNA is the basic building block for all life. When your parents conceived you, they each provided a pre-packaged parcel of DNA that combined to form you. We were told that these special genetic packages were like a living heirloom that was passed from one generation to the next generation without alteration of the basic code. Whatever bits of code you might have inherited from your great-great-great-grandmother were passed on without change.

What a surprise then to learn that the genetic heirloom you inherited may contain some new genetic material depending on what your father ate, or what chemicals he inhaled just prior to your conception!

Over the last 50 years, scientists have also uncovered much more evidence to help us understand the 'fluid' nature of our genes: it has involved the discovery of a network of information pathways linking our genes to the environment. A diagram showing the main information pathways is provided in Figure 1 in the Appendix. You can use this as a map as we navigate our way through the main information pathways linking the environment to our genes. As each new pathway and the molecular processes involved have been discovered, we have been able to forge a greater understanding of the nature of acquired inheritance mechanisms.

WHY DNA ALONE DOESN'T TELL US MUCH

The human genome, the biological data, if you like, that makes us not trees, not sardines, not any other possible life forms, but human beings, consists of over six billion base pairs packed into forty six separately identifiable molecules called chromosomes. Each chromosome is a large molecule containing thousands of genes. If the nuclear DNA is removed from a single human cell nucleus and stretched out to form a continuous chain of base sequences, then it would be several meters long. In school, we were taught that this long sequence of molecules determines what each of us looks like.

So when the idea of the Human Genome Project was first conceived by James Watson in the late 1980s, it was widely believed that what we'd be getting would be a copy of the specs for building a human; a map of ourselves. The only variations that would show up from one 'map' to another would be due to random genetic changes. But the genetic blueprint would remain more or less constant throughout our lives. This was, of course, because we still believed that our genes contained static information stably inherited from one generation to the next.

How distant that all seems now. Still, a genome 'blueprint' was a key objective when the United State's Government's National Institute of Health (NIH) and the British Government decided to provide the billions of dollars of research funds required to support the project.

At first, progress was slow. The project required the coordination of sequencing work conducted by more than thirteen thousand scientists from around the world. However, as sequencing technologies continued to improve towards the end of the 1990s, the scientists were able to dramatically accelerate the speed of sequencing. In 1998, American inventor Craig Venter came up with a new rapid sequencing technology that enabled the Human Genome Project to achieve its goal in 13 years. This was far ahead of the original schedule.

So in 2001, the first map of the human genome was finally completed—a major milestone in the history of genetics. By this time the cost of sequencing was falling rapidly. The result was a rapid increase in the amount of sequencing data available for analysis. A new era had dawned.

Recently, just a few scientists were able to sequence the six billion base pairs of James Watson's DNA in a few months. Watson was chosen in recognition of his crucial role in discovering the DNA double helix, along with the late Francis Crick. Yet, when he was given the data, surprisingly little information or significance could be attached to it. Even at a project cost of an estimated \$US1.5 million, looking at the data it is impossible to tell something as simple as his height or his disease profile with any accuracy. Reportedly, as part of the counselling Watson received, he was advised on the meaning of twenty different mutations that are known to be associated with an increased risk of disease.[43] Scientists can now access the whole genome of many different people. But it's not possible to make medically reliable predictions from any of these individual sequences. They just indicate possibilities.

In fact, the DNA sequence information alone reveals little without also understanding just how fluid our genome is, and how RNA and many other molecular structures function. So just how fluid is our genome?

While DNA forms a double helix in its most stable state, further supercoiling occurs to make the DNA extremely compacted so that the whole sequence is condensed into a tiny nucleus that is only five to ten microns across. This is about one percent of a grain of sand on the beach. Higher order folding also occurs to enhance compaction. Every cell has enzymes whose sole purpose is to unwind the DNA supercoils so that other molecules can gain access to it when required.

Every single human cell also contains the molecular machinery required to fold, repack, cut, paste, copy, delete, insert, store and reshape our DNA in the nucleus. In the process, whole sections of DNA may be deleted, added, broken, rearranged or substituted.

There is also a range of DNA repair and editing machinery on standby at all times waiting to be called into action. For example, if you are becoming sunburnt, the ultraviolet (UV) radiation from the sun produces free radicals in your skin. The highly reactive chemicals produced can result in as many as one million DNA code errors per cell per day! When this occurs, the DNA repair and editing machinery is called into action. It remains constantly active as it responds to DNA damage. There is molecular machinery for repairing nicks, and for identifying and repairing mismatched pairs. As DNA is double stranded, it has the ability to cross-check across a strand and to repair any differences. In the case of DNA repair, if an 'A' replaces a 'C', the 'editor' enzyme detects the presence of a 'G' on the complimentary strand. It then removes the 'A' and reinserts the correct base 'C'. Remember 'C' pairs with 'G', and 'A' pairs with 'T'.

Chemical damage may also cause a physical kink in a DNA strand that can be detected. It is 'snipped' out, and then used to re-synthesize a corrected copy that is reinserted back into the DNA sequence so that the kink is removed. All this is constantly occurring—while you are sleeping, pondering what to cook for dinner, or reading. It's happening right now.

Using long data sequences from mammals, including mice, monkeys and humans, scientists are discovering that our DNA architecture and many sections of our DNA are quite fluid in nature.[44] Some changes occur during cell production, and others are reversible. All processes involving the introduction of changes to our DNA require the presence of a large number of other molecules. Our health and our lifespan depend on our body's ability to actively manage the integrity of our DNA as it undergoes these changes.

It is also now clear that some variable sections of our genome are highly reconfigurable. Yet, we do not understand the nature of many of the mechanisms involved. Why are our genes and genomic structures so fluid? We've had to let go of the old ideas about DNA being a set package. But we haven't yet figured out many of the links between our genes and the myriad of molecular structures that surround them. As fast as things are moving now in genetics, it will probably still be several decades before we understand the fluid nature of our genome or all of the relationships between our genes and their environment. The first link between our genome and the environment involves the production of RNA from our DNA.

MAKING RNA FROM DNA

Next to DNA replication during cell division, the most important job DNA performs is the production of RNA using a copying mechanism known as 'transcription'. Information flows from DNA to RNA.

Transcription takes place in the nucleus and begins with the DNA unwinding. A molecular complex (or transcription factory) binds to the start site, and as it does, a number of molecules assist with the production of a new RNA molecule. Imagine you only have half of a zipper in a sweater, and as you pull the zipper through the zipper fastener (or transcription factory), a new zipper that compliments the original zipper is produced!

During transcription, DNA bases A, G and C translate to A, G and C respectively in the new RNA strand, while Ts are replaced by a 'U' (uracil). For example, a DNA base sequence of AGCCTTT becomes an RNA base sequence of UCGGAAA (since A pairs with U, and G pairs with C). The presence of a T in DNA, or a U in RNA helps the other molecules to tell the difference between our DNA and RNA.

The process of transcription zipper action can occur surprisingly fast, with around one thousand bases per second being copied. There is also strong evidence to suggest that the RNAs themselves play an important role in regulating the process of transcription. In some cases, just a few molecules may be enough to trigger an intricate cascade of transcription events. Zinc ions alone in solution can act as a good catalyst for RNA synthesis. This is probably why many enzymes have zinc ions in their active sites, and why zinc is such an important trace element in our diet.

The result is that many thousands of different types of RNA are being formed every moment of our existence. Digesting food, breathing, thinking and moving all depend on the production of RNAs in exactly the right amounts, in the right locations and at the right time. Some of the RNA informational molecules formed by transcription have a catalytic role. They can cut and splice themselves ('edit') and they are also potentially capable of self replication. Some perform regulatory functions and others are used as a template to produce protein molecules.

After the Human Genome Project was completed, the Encyclopaedia of DNA Elements (ENCODE) project was established to characterise and interpret the complex relationships that exist between our DNA and the RNAs. It was initiated by a public research consortium in the United States in late 2003 to attempt to identify the regions of DNA that are transcribed into RNA by collecting, analysing and cataloguing massive numbers of DNA elements. However, at this stage it is raising more questions than it is answering as no one predicted the complexity of the network of molecular relationships that exist at this level. Our global telecommunications networks are far simpler to understand.

MAKING PROTEIN FROM mRNA

Messenger RNA (or mRNA) is the intermediary involved in the formation of protein structures. Any RNA which acts as a template for protein synthesis is known as 'mRNA'. All other RNAs are known as regulatory RNAs. When mRNA was discovered in the early 1960s, it was soon realised that mRNA was the missing link between our genes and the formation of protein. The synthesis of a protein from an mRNA template is known as translation.

When the flow of information from DNA to RNA to protein structures first became apparent, Francis Crick speculated that the nucleic acid sequences in DNA and then mRNA code was used to form the triplet codes for protein chains. He conducted some initial experiments to show this, and by 1965 we knew that the sequence of nucleotides in DNA corresponded exactly to the corresponding sequence of amino acids. That is, if we identify what the triplet codes for a protein chain is, these can be used to determine the sequence of the mRNA template used to produce it. When this discovery was made, scientists were able to identify the DNA base sequence, the mRNA base sequence and the corresponding amino acid chain produced for all protein molecules. A table showing the list of all possible triplet base combinations and the resulting amino acids is provided in Table 1 in the Appendix.

The process of translation is like threading a carefully copied sequence of coloured beads onto a long string. A single triplet code forms a single amino acid (e.g.

GGA forms glycine). A long string of amino acids forms a protein chain. For example, AUG-GGU-GUA-AUU-GUG-UGA is translated into a START-Glycine-Valine-Isoleucine-Valine-STOP protein chain. If scientists are given a DNA sequence, they can therefore tell what the corresponding mRNA sequence is, and the resulting protein chain structure.

Cracking the translation code linking DNA sequences to the expression of specific protein structures was one of the most significant twentieth century breakthroughs in molecular biology. It has enabled us to develop our understanding of how DNA is translated into a living organism, as well as the study of evolution. Understanding the steps involved in translation has been crucial for scientists developing new diagnostic tests, medicines, and treatments based on gene therapy.

Once translation is complete, there are some additional steps involved to enable our bodies to produce over five hundred thousand different kinds of proteins. Once formed, the protein chain is then compacted into a unique three-dimensional protein structure. Using just twenty amino acids in different combinations and lengths, protein chains can be folded into an almost infinite variety of possible three-dimensional protein structures.

Like mRNA production, proteins are produced in the right quantities, at the right time, and in the right location in a highly coordinated manner. If for example, energy is to be taken from a chocolate bar in your stomach, each step requires a specific protein molecule called an enzyme that has been purpose-built to perform one function. The general rule is: 1 mRNA = 1 ENZYME = 1 ACTION.[45] Altogether hundreds of enzymes are required. One signals that an energy source is in contact with the cell membrane, another extracts the glucose molecules, another allows the glucose molecules to enter the cell, and then others initiate further reactions to progressively strip the glucose of its energy by breaking the chemical bonds binding its atoms together. One by one as each chemical bond is broken, the energy is released for use by the cell. There may be thirty or more enzymes on standby just to identify when a suitable energy source is in contact with the cell surface. This requires a whole army of different enzymes to be on standby. The cell coordinates and regulates all of the enzymes and the processes involved in making energy available as it is required.

Other proteins are needed to respond to our environment. For example, the demand for new proteins such as adrenalin, hormones or muscle power can come suddenly. If you spot a King Cobra near your foot, you are able to respond almost instantly and without thinking. In this case, as the image of the King Cobra is processed by your brain, it triggers an automatic nervous system response involving several different pathways. Your heart beats fast, your palms sweat and your muscles involuntarily contract. Activation of all of these reactions relies on the central nervous system and the extremely rapid translation of proteins and hormones to enable you to mount a fully coordinated response. Once you are out of danger, the response is terminated.

A similar but slower translation response is initiated when a child's permanent tooth is knocked out. When the tooth is placed back into the gum, its base makes contact with the child's bloodstream. This contact triggers a series of translation events to enable the tooth to rapidly regrow and re-anchor. The mending process is terminated when the tooth has re-established itself firmly in the gum. Each step involved has to be initiated, and a signal is required to terminate the process of reanchoring the tooth. Each individual reaction requires the translation of one or more enzymes. However, the process fails if the tooth is not placed quickly back into the child's gum before the wound starts to close.

Genetic translation processes like these are essential to all life. For example, when the silk gland of a silkworm larva produces its sought after material, it does so by simultaneously translating the long threads of fine silk protein from mRNAs. In nursing her newborn, a mother triggers action in a wide range of proteins required to produce milk.

There are also some more unusual examples involving transcription.

The salmonella bacterium has a number of flagellae, 'tails' that are rotated to move it along like a propeller. Each flagellum is a single strand of protein that can switch between two distinct proteins, caused by the periodic inversion of a small segment of DNA. That is, salmonella bacteria are designed to periodically alter the type of protein that they produce from mRNAs in order to move. Why? Well rapid changes make it harder for our immune system to figure out what it is up against, which makes it hard for the body to come up with an effective response. Unsure which way to jump, the immune system can't help us, and we can become very ill. This is another example where we continue to be outsmarted by a single celled life form.

As a species, the Death Cap mushroom that grows throughout Europe has evolved a different type of transcription-based strategy. The Death Cap mushroom is believed to have killed Pope Clement VII in 1534, and the Roman Emperor Claudius in AD 54. It wards off predators by interfering with the process of transcription. It produces a chemical that disrupts the formation of the mRNAs needed to produce proteins in humans. About three days after eating the deadly Death Cap mushroom, humans suffer severe abdominal pain. This is quickly followed by liver and kidney damage, and seizures. If untreated, up to seventy percent of people will die—hence the name. Any survivors are unlikely to go back for a second helping, so the mushroom protects itself.

There can be some quite unexpected outcomes if errors are introduced into the protein folding patterns during the post-transcriptional phase. Over the last few decades, some of the relationships between protein misfolding and disease have been discovered. In bovine spongiform encephalopathy (BSE), better known as 'mad cow disease', we have found that the trigger is conformational change rather than a nucleic acid sequence change.[46] In other words, there is some other agent responsible for causing the protein misfolding. We still do not know what it is, which means we can't yet develop a cure. The disease Kuru is also known to be linked to protein misfolding. Kuru is an infectious brain disease affecting highlanders in New Guinea who practice cannibalism. There are some similarities between the symptoms exhibited by sufferers of Kuru, Alzheimer's disease and Parkinson's, which affects millions of people worldwide. All of these are linked to protein misfolding. But the source of the infectious agent for any of these fatal neurodegenerative diseases is still unknown. While various agents such as 'prions' have been hypothesised, we still do not fully understand the molecular mechanisms involved.

We do know that translation and transcription are temperature dependent processes. In the laboratory, if the temperature is below 25°C it becomes extremely difficult to separate DNA into two single strands. This makes it impossible to produce the single stranded DNA template needed to actively produce the mRNAs for protein synthesis. When the temperature rises above around 75°C, it becomes impossible for DNA or mRNA to form stable bonds with complimentary strands, therefore making it difficult for complex metabolic processes involving strong pairing between molecules to occur.

This is probably why we have evolved as warm blooded animals—to ensure that we remain metabolically active throughout our whole life. The enzymes required to enable us to keep breathing, to keep our temperature constant, and our hearts beating even when we are in a deep sleep mean transcriptional and translational activity never ceases. But with cold blooded animals, if the external temperature drops low enough, the core body temperature of the animal falls too. As it does, it becomes harder to maintain the transcriptional and translational metabolic processes required to ensure the animal remains metabolically active, and it hibernates. Once again, when the weather becomes warm, the animal slowly becomes more metabolically active.

Yet, there are some exceptions. Some types of fish in very cold water do not need to hibernate. We do not fully understand why they can remain metabolically active in temperatures that are close to the freezing point of water. Some strains of bacteria have been able to evolve mechanisms enabling them to adapt to live in very low temperatures at or below the freezing point of water—or at extremely high temperatures such as in a hot spring. There are also animals that can remain in a metabolically inactive state for several years, while others have evolved to leap into metabolic activity only when water is present.

The process of transcription is also sometimes prevented by active suppression. For example, during sex several hormones are produced to suppress normal transcription activities. This improves our 'efficiency' by freeing up the resources for the massive hormonal release during sex. As a blood-stirring set of reactions are initiated, the hormones released during sex travel on specific carrier proteins to their point of release in target tissue. In the target tissue, the hormone triggers the translation of proteins. The whole sex act becomes the focus of inputs from all of our senses, including touch, taste, sound, sight and smell. In a moment of passion, the massive hormonal flow also triggers an increase in our heart rate and breathing.

Thus every moment of our lives we are totally dependent upon a wide range of translation events occurring in a carefully coordinated and synchronised manner. Together, these processes enable us to function and rapidly respond to unforseen events.

THE UNIVERSE OF RNAs

As the known number of RNAs has increased, so too has our understanding of the RNA regulatory world and the role of RNAs in evolution. The regulatory RNAs are all of the RNAs that are not involved in the formation of protein structures. For this reason they are sometimes referred to as the 'non-coding' RNAs.

For most of the last fifty years we thought that RNA was simply the messenger between our DNA, held securely in the nucleus, and the cell where the protein production machinery awaits. Over the last decade or so, scientists have begun to find all sorts of other roles for RNAs. These have given us a much broader understanding about the complex processes underlying the rather fluid interplay between the genetic elements of nature and nurture.

In fact, we have now identified as many as thirty seven thousand new RNAs in the human genome, each with a different prefix. Some RNAs regulate the levels of expression of different proteins. Others edit other RNAs. Many are organ or tissue specific, and some play a key role in managing gene expression. RNA to RNA, RNA to cell, cell to cell, cell to organ, organ to organ or organism to organism signalling functions all rely on RNAs. To enable humans to communicate with each other requires a whole orchestra of RNA activity to be conducted by us.

In other cases, RNAs play a lead role in chemically altering the surface of our DNA. They perform the very neat trick of chemically altering the DNA surface structure, without altering the DNA coding sequence. The chemical changes to the surface structure of DNA are known as epigenetic markers as they literally 'sit above the gene'. The resulting changes give rise to some amazing diversity among individuals in a population. That is, epigenetic markers can change the expression of our genes so that two individuals with identical DNA sequences can appear and act differently. These discoveries are highly significant. They show that we not only inherit DNA from our parents to define who we are, we also inherit a whole range of epigenetic markers. This provides another mechanism whereby acquired changes can be inherited. We look at some absolutely fascinating examples of acquired inheritance effects involving epigenetic markers in Chapters six and seven.

RNAs are also providing scientists with a new source of explanations for illness. Many RNAs are linked to cancers, diseases of the central nervous system, and common infections. New medicines that target particular RNAs, and shut them down, could help cure some of the RNAs causing disease. By shutting down a particular RNA, the production of viruses or rogue protein cells can be stopped. This can be done by targeting the section of DNA responsible for transcribing the RNA. Or it can be done by targeting a specific RNA. A small complementary sequence of only around twelve to twenty five nucleotides can be joined onto the target sequence, thus preventing it from performing its normal functions, or even dicing it to pieces.

We have only recently begun to explore the universe of small RNAs and how they function. Understanding their roles in cells, in specific tissues, and at the whole organism level will also be crucial. The consequences for medicine and our understanding of acquired inheritance effects are profound.

In the process of discovering more about the universe of RNAs, we are also discovering that it doesn't really matter how much DNA sequencing we do. That alone is not going to reveal a great deal of information about us, or how we function as individuals. The massive number of regulatory RNAs, and protein coding mRNAs also define who we are. The complex and dynamic patterns made possible by the RNAs are capable of producing extraordinary diversity well beyond what can be generated by our DNA alone.

The sheer complexity and complicity of the new universe of RNAs also means that we have to redefine what 'hereditary material' is. We are in the midst of a paradigm shift: we are leaving behind the view that genes are a stable repository of information used for building proteins. While all RNAs are transcribed from DNA, the picture now emerging is of a far more sophisticated universe of RNAs. The expanding universe of RNAs consists of a complex network of cooperative molecular elements responsible for a wide range of functions.

From an evolutionary perspective, recent discoveries make it plain that our evolutionary developments are far more about changes in the regulatory RNAs and the genes that do not code for proteins, than about random errors accumulated in our DNA.

THE CELL FACTORY

The human body consists of around one hundred trillion cells, organised in a precise and carefully managed assembly of bones, organs, muscles and fluids. The development and maintenance of such a large assembly of cells is an extraordinary triumph. But how is this triumph achieved? What role does the cell play in the information flow linking the environment to our genome?

A single cell is only around 10 to 20 microns across, but it is an amazingly complex thing. Think of it as a tiny self-powered factory. It is protected by a cell membrane and has a range of protruding environmental sensors. It has a nucleus of around 5 to 10 microns at its centre to protect the DNA, and the fluid inside the cell that contains numerous molecular structures. Cell to cell signalling enables cells to communicate with daughter cells, and with other types of cells. It enables cells to work cooperatively with rules and controls, and to differentiate themselves from nearby neighbours when required.

At a molecular level, a typical cell is a highly intricate structure consisting of around ten million million atoms. If we magnified it, it would resemble a highly flexible and automated factory consisting of several different production lines, a control room, small robots for responding to instructions, and sensors to ensure that the factory inputs, outputs and processes all work in an efficient and coordinated way. The control room collects and integrates a range of signals. It regularly monitors quality and corrects mistakes in all areas. It can alter the processes to improve quality and efficiency. Sometimes it changes the processes involved to meet new requirements. It has its own inbuilt repair team on standby, ready to identify and repair any damage. Yet it is also capable of duplicating itself, and responding rapidly to changes in the external environment. In reality, the cell has all of the machinery on board to ensure that chance mutations are rare events indeed.

There are hundreds, possibly thousands of different types of surface receptors that enable the cell to sense and respond to environmental stimuli. They include heat sensors, shapes for detecting the difference between the cells of the host and those that are foreign, pressure sensors, light sensors, and many more. All this enables cells to act as secure gatekeepers, allowing for controlled entry and exit of a particular type of molecule through the cell membrane. As the gatekeeper between the environment and the protected nuclear DNA, each cell takes in a wide range of information from the external environment using the sensors fitted to its surface. It processes the information. Sometimes the cell is triggered to reproduce. Sometimes it is required to inject more energy, or to expel unwanted waste, toxins or degraded foreign nucleic acid and proteins. It has its own memory to know what molecular structures to produce, when to produce them, and how many are required. It knows how to update its RNA or DNA. The inner fluid of the cell is continually communicating with the DNA in the nucleus, and with the cell membrane that faces the external environment.

Using translation processes, proteins are the main product manufactured by the cell "factory"—they take up around ninety percent of the total amount of chemical energy used by a cell. Each cell contains several thousand copies of many different types of proteins and RNAs. The translation machinery calls on over two hundred different types of proteins and enzymes, and more than forty different types of RNA. Overall, more than three hundred different macromolecules are required for protein synthesis, and all are made at a very fast rate. The process is also highly regulated to ensure that just enough copies are made of each, and that the right ones are available at each stage in the lifecycle of a cell.

In a rapidly changing metabolic environment, maintaining a healthy balance between protein production and the breakdown of protein is extremely important. An inability to reduce the number of these proteins can result in uncontrolled tumour growth. But if the available proteins break down too rapidly, the result can be renal disease, asthma or a neurodegenerative disease like Parkinson's disease. Specific proteins are involved in specific diseases. For example, cystic fibrosis is the result of too rapid a breakdown of an ion channel protein. The results for sufferers are catastrophic.

When a cell replicates, a huge number of molecules are involved. In *E. Coli* bacteria, at least nine different proteins are involved in the initiation phase of the replication process alone. Replication in higher cells is even more complex. In humans, some cancers develop when damage occurs to some of the many genes involved in normal cell replication. Yet at each highly regulated stage, there is error correction, and the complex pathways have a lot of compensatory mechanisms built in. In the human genome, we have identified more than one hundred and thirty genes that encode proteins dedicated to the error detection and repair process alone. Many of these are integrated with the DNA replication system. Even if one gene is missing, or unable to function properly, cellular replication can often proceed normally. The cellular replication process also involves a lot of unusual genetic re-arrangements, although the purpose for these is not yet known.

As there are a large number of error detection and repair mechanisms involved, a lot of genetic damage needs to be accumulated before most diseases can progress. In the case of nonpolyposis colon cancer, damage to at least five different genes must occur before the disease develops. Some breast cancers are the result of damage to two or more genes.

A lot of cellular gene damage can be directly repaired. However, when there is a lot of damage, sometimes SOS proteins are called into action to create new protein. SOS proteins enable the other molecular machinery to replicate past many of the DNA lesions that would normally block replication. When DNA is extensively damaged—by UV exposure for example—DNA replication is halted and proteins bind to the damaged area. This is what happens when a small melanoma starts to develop. The body mounts an SOS response to protect the rest of the body. At the same time a number of other DNA repair, excision or inactivation actions are initiated. The SOS proteins then establish a communications link between the signals indicating the area where DNA damage has occurred, and they coordinate the SOS response. Mounting an SOS response involves a lot of cross-gene coordination.

Cells are also capable of changing their form to meet altered environmental conditions. In the case of the receptors included on the cell surface, some are used more often than others. When a particular type of receptor is needed, its numbers increase on the surface, and other types of receptors no longer needed die. The next generation of cells maintain the new receptor levels. They have adapted to the environment and passed on their new characteristics. In this sense, acquired inheritance effects are operating at the cellular level.

Although all of the activities of a single cell are highly regulated, cell membranes are sometimes 'tricked' into letting invading viruses enter and make use of the cell's molecular machinery. All viruses are simply made up of small strands of DNA or RNA that are protected by a protein or membrane coat. However, they cannot grow or multiply without entering a host cell: they need to hijack the internal cellular machinery of a cell to reproduce and ensure their own survival. When a virus first infects us, it enters a cell, uncoats its genome, and then proceeds to take over the cell's metabolic pathways for its own benefit. Viral RNA or DNA strands are then replicated using the cell's molecular machinery. The virus is also able to make new protein coats from its DNA or RNA template inside the cell, as well as produce viral progeny that go on to infect other cells. In this sense, viruses are the ultimate parasites of the cellular world.

There is another nifty trick in store, too. Once in a host cell, foreign viral RNA or DNA can mutate. Flu viruses mutate so rapidly that the host's immune system can be evaded for weeks. In the case of the flu virus, the host cell is killed as the virus replicates, causing us to become very ill. Yet other viruses co-exist peacefully. Some, such as herpes, can lay dormant for a long time before replicating.

Another group of viruses are called the endogenous retroviruses. These are the ones already encoded in our DNA in cells. They have standard viral structural features, only they are encoded in the germline. They replicate and transcribe via an RNA intermediate. They are spontaneously produced throughout life from normal transcription activity, and they occur in extremely high concentrations in placental tissue and seminal fluid. While we have known that endogenous retroviruses have existed for more than three decades, their purpose is still poorly understood.

In summary, the cell is designed to enable our genome to interact with many types of molecules and the external environment. The inside of a cell can be viewed as an RNA encoding factory that is able to selectively pass information across the cell membrane, and to process a wide range of sensory information from the environment. It takes instructions from the environment and it changes itself in response to environmental signals. The cell also feeds information back into the environment in the form of the many types of RNAs and proteins it exports to other cells.

Recent work has also shown that in multi-cellular organisms, the memory trace created by cells as they feed information into and out of the environment is stored as a modification of the connectivity between cells. In bacteria, the changes induced are known to persist for several generations.

There are also other background regulatory processes involved. Some of these help to ensure that the genetic and epigenetic trajectories of ageing, differentiation and reproduction are supported by a rapid and highly complex network of RNA production sites. Others include managing the cell to cell signalling processes that are required to provide spatial and temporal feedback, or organ to organ feedback. This is similar to the signal inputs that we are required to know when we need to drink, what we need to drink, and where the glass is. These are important for managing and coordinating overall status updates for cells, organs and our own body.

UPDATING OUR DNA

Back in the late 1950s, US geneticist Howard Temin had noticed that RNA containing tumour viruses simply disappeared when he added them to tissue cells grown in the laboratory. At the time he could not explain this result. Eventually, he speculated that the virus had copied its RNA into the tissue cell's nuclear DNA by a process called 'reverse transcription'. That is, he predicted that genetic information flows from RNA to DNA—the opposite to what most other scientists believed at the time. Information flow from RNA to DNA is called 'reverse transcription'.

Temin also reasoned that there must be an enzyme present to copy the RNA into the DNA. Making this prediction also took courage, given that the great Francis Crick had propounded what is known as the 'Central Dogma' of biology. In it, he stated that information flows from DNA, to RNA and then to protein, 'NEVER the reverse'. Temin therefore had few supporters at that time.

It wasn't until 1970 that Howard Temin was able to demonstrate in experiments that tumour cell viruses from mice and chickens broke this cardinal rule. He showed that RNA, including some cancerous RNA viruses, could be reverse transcribed into the DNA. Any enzyme that can make a copy of an RNA template, and copy this into the DNA is now called a reverse transcriptase. In separate papers appearing in the same issue of *Nature*, both Howard Temin and his compatriot David Baltimore[47] independently verified the existence of the reverse transcriptase enzyme responsible for reverse transcription that was predicted earlier by Howard Temin. In 1975 Howard Temin and David Baltimore were awarded the Nobel Prize for this work. Their experiments were important in that they showed that retroviruses from the environment can enter and attack a host. The viruses enter a host cell where DNA copies of the virus RNA are produced. The DNA copy of the virus is then inserted into the DNA in the cell nucleus in a process called reverse transcription. The updated cell then goes on to reproduce "daughter cells" with the extra DNA cargo when the cells multiply. Some types of breast cancer are the result of a type of retrovirus that becomes integrated into the DNA of the host cells.

So by the mid-1970s, the new age of genetics involving the use of recombinant DNA technologies had arrived. By this time, scientists had discovered overlapping genes, genes within genes, genes that can code for more than one protein, and many other genetic phenomena. Then in a major development, Americans Stanley Cohen, Herbert Boyer and Paul Berg successfully inserted a gene from an African clawed toad into bacterial DNA. There were also reports that suggested transcription can start at one gene and move to another to make a particular protein—called a 'fused transcript'. In the laboratory, genes could be cloned, techniques for sequencing were becoming widely used, and new genes could be synthetically sequenced from scratch in the laboratory. DNA sequences could now be cut and pasted at will. It was also the decade in which scientists started to develop the first transgenic animals referred to in Chapter 2. With so many new technologies becoming available, it was soon conclusively shown that non-viral reverse transcription is widespread in our genome.

By the late 1980s the first testable molecular models explaining how reverse transcription occurs were proposed. However, debate continues about which model is correct. Thanks to this research, we now know a lot more about 'infectious retroviruses' that adopt a very deadly strategy. Infectious retroviruses make a DNA copy of their RNA template, and then insert it into the host cell's DNA in the nucleus. It is a more deadly strategy because the cells of the host organism then continue to propagate the new infectious DNA every time a new cell is produced as a part of normal cell division. Once this occurs, new RNA copies of the virus are made in the new cells, thus allowing the infectious retrovirus to be active and cause havoc in many new cells throughout the body.

This is the strategy adopted by the infectious retrovirus known as AIDS (or Acquired Immune Deficiency Syndrome). After the AIDS RNA virus enters the body through sex, contaminated needles or a blood transfusion, it goes to its favourite host, the T cell. The AIDS virus genome consists of a large single strand of RNA that is nine thousand nucleotides long. When it replicates, it is around ten times more error prone than other retroviruses. Its rapid rate of mutation enables it to continually evade any defence that the human body can develop against it. It also means that several strains of the AIDS virus can exist in a single host. This makes it impossible to develop an effective vaccine as the AIDS virus is replicated freely by the host and remains hidden from the immune response. The process becomes difficult to stop, especially if the infectious retrovirus mutates very rapidly. This is similar to a strategy used by some trojan horse viruses that infect computers. Other examples of infectious retroviruses include some types of breast cancer and leukaemia. When a retrovirus is taken up by a new host, it becomes integrated into the DNA of the host by the reverse transcription mechanism. Some are airborne. Some are eaten and taken up.

A second group of reverse transcriptases occur naturally in our genome. These are the endogenous retroviruses. We all have hundreds of thousands of these in our DNA. We do not know the function or the origin of most of these. Some could have originated from an external source at some stage in our evolutionary past.

Some endogenous retroviruses can cross the species barrier. We do not know how many are active, and we do not know what triggers activate some latent forms of endogenous retrovirus. Some transfer also occurs when any two individuals exchange body fluids such as saliva or seminal fluid. Some viruses may lay dormant in the new host for decades before springing into action, and others may be slow acting, such as the human papilloma virus (HPV). Thus, how often we exchange bodily fluids will potentially have a huge impact on our own individual retroviral load.

Evidence is mounting that another group of reverse transcriptases are involved in the repair of lesions in our DNA.[48] When there is no template for the repair mechanism to work from, then one of the safest options is to use a reverse transcriptase to copy and paste in a section of an RNA sequence that is recognized as self and which the body thinks is harmless.

In most of our cells, chromosomes have telomeres at their ends. Structurally, each telomere is a long stretch of DNA at the end of a chromosome that contains a series of repeat patterns of DNA base sequences. In 2009 Elizabeth Blackburn was awarded the Noble Prize in Physiology or Medicine for her work on telomeres. She discovered that the enzyme telomerase which is a reverse transcriptase, and a tiny RNA consisting of no more than six to twelve nucleotides, together play a key role in synthesizing a telomere by copying, and pasting the repeat small DNA sequence into the telomere.[49] Just like in a tiny sausage factory, the telomerase and the tiny RNA sequence repeat the process. One hundred or more extra copies of a tiny repeat sequence may be added. The repeat base sequences are also species specific. The telomeres at the tips of each of our chromosomes in tissue cells shorten each time the cell replicates. Finally a point is reached where the cell is unable to divide. This process occurs in all cellular DNA after about fifty replications. However, we are still not clear on how this relates to our own aging processes. Yet oddly, most organisms also have the molecular mechanisms needed to extend the length of their chromosomes. We have also discovered that the telomeres become fragmented during normal cell division, and we are not sure why.

Andrea Bodnar, working from the Geron Corp. in California showed that lengthening the telomeres of chromosomes in cells in culture could extend the life of cells by more than 20 divisions compared to normal cells.[50] However, it is still unclear if we know enough to be able to translate these findings to extend our own natural life span.

It is possible that these processes may somehow be involved in providing additional space in our genome ready for new DNA sequence information to be written into the DNA using reverse transcription. The repeated DNA sequence in a telomere of say,

'TTAGGG TTAGGG TTAGGG TTAGGG'

might be decoded by other molecular machinery to read, "this space is intentionally left blank". This is like buying a 500 sheet ream of blank A4 paper sheets ready for printing. Each sheet is exactly the same size, colour and weight. We place some of the blank sheets in our printer ready to print a hard copy of new information that we wish to retain. Sometimes we need 15 sheets, and at other times we might only require three. While this imagery might be somewhat simplistic, further experiments will shed more light on the mechanisms involved.

In all known situations where reverse transcription occurs, the process relies on the availability of an RNA template to be copied. With the knowledge that there are also several different types of reverse transcriptases able to be recruited for reverse transcription, then it is likely that it has played a key role in directing our evolution. The reverse transcription processes involved provide huge potential for updating our DNA during our lifetime.

A MISSING LINK?

Reverse translation is a speculative process involving the transfer of information encoded in the form of protein into RNA or DNA. I say speculative because when we look at a map of the known information flow paths (in Appendix 1, Figure 1) between our genes and the environment, the process of reverse translation is a highly conspicuous missing link. We have no evidence to suggest that information in the form of a protein can be transferred directly to a nucleic acid.

It seems that with so many other RNA controlled regulatory and feedback mechanisms involved in delivering information to and from our genes, the direct information flow from protein to DNA or RNA is not required. Nor is there any direct evidence suggesting that information is transferred directly from protein to protein. These are views that most biologists subscribe to today.[51]

REARRANGING OUR GENES

The process of genome-wide DNA rearrangement is known to occur in many higher animals and single-celled organisms.

In the single-celled protozoa *Oxytricha,* the whole genome undergoes a massive re-arrangement.[52] During the rearrangement ninety five percent of the genome is eliminated. The remaining few short strands of the genome are then rearranged and some telomeres are added to extend the length of the chromosomes. Finally, amplification processes produce an altered version of the genome.

In many multi-cellular organisms, the process is quite perplexing. As the details are unravelled by scientists, we are discovering a whole new range of mechanisms that make it possible to rearrange the genome each time a cell divides.

What is most remarkable about the process of updating and reassembly of a genome is that each time it occurs the same precise locations on each chromosome are used as rearrangement junctions. This is like shuffling a pack of cards and ending up with exactly the same order of cards that you started with. Imagine if we were able to quickly do this with a 3 million card deck. This is really amazing—and quite impossible to account for in terms of incremental cumulative selection! We also now know that the RNAs play an important role. Some RNAs are used as a type of 'LOOK UP' table to direct the restructuring of the genome.

While we still do not know how widespread RNA-guided rearrangement is, these recent findings have some clear implications for our understanding of how our genes function: they provide further evidence of the fluidity of our genes, and the processes that enable new somatic information to become integrated into germline DNA. It will be interesting to learn which parts of the human genome are affected in humans.

OUR GENES AS A 'LIVING CODE'

It is becoming increasingly evident that nature has provided us with a highly sophisticated 'living code' that is capable of coordinating the processing of millions, or even billions of instructions simultaneously. The network of fixed and mobile molecular structures can be viewed as a highly regulated system with inbuilt redundancy. It has its own error checking, environmental sensors, and parallel processing powers to perform many functions simultaneously. It also has the ability to alter itself to meet new environmental challenges. The function of our genes can therefore only be understood in the context of a whole organism and the environment.

It is also evident that different sections of our genes are able to evolve at different rates.

The twenty thousand or so protein coding genes are almost identical throughout the animal kingdom. From the worms, through to the squids, reptiles, birds and mammals, the protein coding genes have remained largely unchanged for over five hundred million years. In bacteria, most of the genes are for coding proteins. This was established in the mid-1970s.

In 1977, it was discovered that the protein coding genes in all higher multi-cellular organisms, including humans, includes large tracts of nucleic acid that do not actually code for protein. These are generally referred to as non-coding regions of our genes. From an evolutionary perspective, Australian genomics researcher John Mattick and his colleagues have shown that as an organism has become more complex, the larger the proportion of non-coding genes in its genome.[53] Thus, from the earliest earth worms to man, one finds that each contain about twenty thousand protein coding genes. But what has happened during the last billion years has been a massive expansion in the proportion of the genome devoted to non-protein coding genes. In earth worms, about forty percent of the genomic DNA includes protein coding genes that are transcribed. In higher mammals, more than ninety eight percent of the genomic DNA is devoted to synthesising non-protein coding RNAs. In humans, the non-protein coding genes represent a huge ninety nine percent of our genome. The conclusion is that most of the enormous number of non-coding genes are required to produce regulatory RNAs for complex living organisms like a human. Although there are also many repeat sequences in the non-coding regions of our genes, it means that most of the human genome could be functional after all. There are also some regulatory RNA transcripts and related sections in the genome that are highly mutable: different rates of mutation and different mechanisms for generating mutations are evident across a wide range of RNAs that do not code for protein. This means that the regulatory RNAs harbour huge potential as a large set of highly mobile and mutable molecular structures. They work cooperatively to enable humans to respond to changes in our environment and to remember successful responses as we live day-by-day.

It is the huge amount of diversity that is possible to be generated by the more mutable elements that has, and will continue to allow us to evolve even greater complexity in the future. At the same time, we need to analyse our human genome and epigenome as a single information management system, rather than as a collection of separate nucleic acids, molecules, and proteins: it implies that feedback exists at all levels to ensure that complex life forms can be sustained and continue to adapt. It also provides the scaffolding upon which acquired inheritance effects can be generated.

* * *

So what does this mean for our future?

We are likely to make many more fascinating discoveries about how we function. But only by taking a more holistic view will we be able to begin to unlock the algorithmic potential that exists among the elements involved in managing a fluid genome. The promise is a whole lot more insight into the complex causal relationships that exist among the protein coding genes, our regulatory genes, our disease profile, and how we might control our future evolution.

The process of discovery will involve amassing huge volumes of data, and an even larger amount of related data. It will require highly sophisticated computers and software for analysis. Even building the analytical software required will be a challenge.

Conversely, as the next generation of code crackers discover more about the algorithms used by a living code, they will probably be able to use these to improve their own ability to conceive and develop more intelligent evolutionary software. By developing self-learning programs that emulate how the genome adapts to its environment, we could start to come close to being able to perform some of the complex computational feats performed within each living cell. If the next generation of code crackers can achieve this goal, then they will be the ones to truly unleash and inherit the computational powers within.

* * *

5. ACQUIRED INHERITANCE EFFECTS IN THE IMMUNE SYSTEM

As we live our lives we are bombarded by armies of new viruses and bacteria. While many of these are harmless, some are not. This challenges our immune system to be able to rapidly mount a response to an almost infinite variety of new foreign pathogens. As some of the more deadly viruses and bacteria can also mutate rapidly, we need to be able to develop a flexible and fast response. As a result our immune system genes are constantly managing a relentless war within.

What is most remarkable about our immune system is that it learns how to fight off previously unknown bacteria and viruses by creating whole armies of new antibodies. It also remembers how to fight some of the most lethal pathogens that our ancestors were exposed to. The processes the immune system relies on are, therefore, Lamarckian in that they require the antibody genes to *adapt* to fight a new foreign infection, and then *pass on* the ability to create more effective antibodies to offspring. To do this, the immune system relies on feedback to our genomic DNA using reverse transcription for the benefit of the next generation: in fact, the nature of infectious diseases that we have been exposed to in our evolutionary past has been the only environmental force guiding the evolution of the vertebrate antibody gene family.

EARLY EXPERIMENTS

The first experiments drawing attention to acquired inheritance effects operating in the endocrine system were conducted on lab animals in the 1960s and the early 1970s. They relied on injuring endocrine organs such as the pancreas to give rise to diabetic rats.

Experiments by K. Okamoto in Japan[54] and later M.G. Goldner and G. Spergel in the United States[55] used the drug Alloxan to experimentally induce diabetes in laboratory rats. Alloxan was known to irreversibly damage the tiny insulin producing islet cells in the pancreas, resulting in very high sugar levels in the host. These experiments showed that after diabetes was induced in rats, the disease was transmitted to successive generations of progeny. They found that diabetes developed in large numbers of first generation progeny. They also found that inbreeding among the diabetic rats caused the disease to become progressively worse in each subsequent generation.

Similar results for acquired inheritance effects were reported by scientists doing thyroid dysfunction experiments. By damaging the thyroids of rats, it was shown that untreated progeny were born with malfunctioning thyroids.[56] What wasn't known at the time was the role of DNA in propagating the organ defects to the next generation. Some autoimmune responses may have been involved during the reactivity.

By the 1970s, scientists started to recognise that there were close links between our DNA and disease. By then, our growing knowledge of the immune system was providing some important clues in our understanding of how updated genetic sequences can be created in order to fight foreign pathogens in our environment. These clues enabled Australian immunologist Ted Steele and his Canadian colleague Reg Gorczynski, to conduct the first acquired inheritance experiments on the immune system.

Steele and Gorczynski began by looking at measurable functional responses to immune system challenges. They asked whether altering the immunological function of the father by active treatment with an antigen could lead to progeny showing the same altered immunological function. Troubled by the conventional neo-Darwinian explanations that strictly forbade any mixing of somatic DNA with germline DNA, Steele and Gorczynski said it was ridiculous for scientists to place some sort of ideological 'chastity belt' around the germline DNA. They predicted that there was no Weismann's barrier, and that some of the new DNA was copied into germline cells ready for the next generation.

To build a theoretical model to explain their predictions, Steele relied on Temin's ideas on reverse transcription as the mechanism to copy new RNA into the DNA. Once new RNA sequences were copied into the DNA of somatic cells, he predicted that it would also be copied into the sperm and ova. When Steele first published this theory on 'Somatic Selection and Adaptive Evolution' in 1979[57] he was acutely aware of the Lamarckian implications, and that it was against what other scientists believed at the time.

Shortly after the publication of his ideas, Steele published the first experimental data with Reg Gorczynski at the Ontario Cancer Institute. They showed that immunological tolerance developed in male mice can be passed from one generation to the next.[58] In these experiments, they mated inbred male mice with an altered immunological function to normal (untreated) inbred females. The experiments were not conducted using immunised females as there is a direct link between the foetus and the mother during gestation and via early milk that could account for any transfer of antibodies or genetic information between a mother and her progeny.

Gorczynski and Steele reported that if newborn male mice of strain A were repeatedly exposed to lymphocytes of strain B using a tissue graft, then aspects of this specific tolerance developed to strain B were observed in progeny born to untreated females. That is, the specific tolerance developed when they were young was passed on to the next generation through the male sperm. Despite some harsh criticism about the experimental approaches used, and conjecture about the mechanisms involved, Steele and Gorczynski believed that they had demonstrated that acquired inheritance effects were real.

Steele continued to conduct acquired inheritance experiments using mice at the Clinical Research Centre in Harrow, London.[59] Negative results from these experiments were published in 1981 by a group of highly respected and more senior scientists working at the Clinical Research Centre in Harrow.[60] But one week later Steele swiftly refuted their claims in *New Scientist*.[61] By analysing their data carefully, he argued that, in fact, it showed some of the acquired inheritance effects he had previously predicted and Steele was again harshly criticised. The dispute ignited a fierce and very public debate that divided the scientific community.

The debate was reignited in 1983 when similar results were reported by another group of scientists working at the Clinical Research Centre in Harrow, London and the John Curtin School of Medical Research in Canberra.[62] Using a different test system involving mice, they reported that immunological effects were passed on to the next generation through the male sperm.

In later experiments, Steele attempted to replicate the acquired inheritance experiments conducted a decade earlier using mutilated endocrine organs. He used inbred mice, and induced diabetes in males using the drug Streptozocin. The experiments had limited success. The main findings reported were abnormal variations in body weight and some developmental defects. Only one spontaneously diabetic mouse was found among the two hundred and thirty five progeny tested.[63] Steele has persisted with his work on acquired inheritance effects to the present, although his work has continued to be basically ignored.[64] Steele described his battle in attempting to make the immunologists more receptive to Lamarckian ideas in the 1980s, thus:

> *"It was a very painful time . . . If I had seen the human mangling machine I was entering, I probably would have said 'I don't want to do it',"*

By the early 1990s, a number of other molecular experiments had been conducted to develop an understanding of how antibody genes are updated over evolutionary timeframes as a result of exposure to new viral and bacterial infections. By this time, it was known that each new infection we are exposed to in our lives leaves behind its own unique somatic signature in our genetic DNA: evidence that we have been exposed to it. Such somatic mutation patterns have now been found to exist in a wide range of vertebrate antibody genes. These simply cannot be explained using a neo-Darwinian explanation. The new DNA sequence can only be explained if DNA has been transferred from the soma to the germ line. Reverse transcription of the new RNA sequences into the DNA is a crucial step.

This suggests that the feedback process used by the immune system to update our repertoire of antibody genes has been active for at least the last four hundred million years. It implies that the antibody genes have evolved using a distinctly Lamarckian mode of inheritance.

By 1998, when we published *Lamarck's Signature*,[65] there was already considerable indirect evidence to suggest that some sections of the antibody genes may be altered by the body's experience, and that this is passed on to the germ cells. At the time, it was believed that the main evidence of a soma to germline transfer of information only applied to cells actively producing antibodies. That is, the Lamarckian modes of inheritance being exhibited by our antibody genes did not apply to any other families of genes. At the time our book was published, the slowly changing view of the biomedical community was summed up by Australian scientist Adrian Gibbs, a specialist in virus evolution, when he said,

> *"The ideas presented in Lamarck's Signature are almost certainly correct. Their critics must produce a better explanation . . ."*

While the use of the immune system genes alone is not sufficient to convince most people that Lamarckian inheritance effects have played an important role in our evolution, the critics have not been able to provide a better explanation. So how did the highly non-random patterns observed in the antibody genes arise?

THE IMMUNE SYSTEM IN ACTION

There are potentially tens of thousands of genes in our DNA that can be called upon to fight infections. The DNA sequences in the variable regions of our antibody genes provide a partial historical record of the range of infections that our forebears were exposed to.

The vertebrate immune system detects infections, it responds to them, and it remembers the encounter. We are all born with our own defence mechanisms, and a 'library' of antibody genes for creating a fast response to the most likely invaders. That is, those that our ancestors were exposed to. This means that infants are born with a ready response to pathogens that their ancestors have had to respond to.

One of the best illustrations of this in action was when Europeans settled in the Unites States, bringing with them smallpox. While most European settlers survived smallpox, it devastated the indigenous populations who had not previously been exposed to it. Similarly, when Europeans settled in Australia, they brought with them the flu virus that destroyed some indigenous communities. All viral infections such as the common cold are intracellular parasites that can only live and reproduce in a living cell by harnessing the cellular machinery for their own benefit. In the process they also mutate and the progeny viruses go on to infect more cells.

In each case the European settlers had the advantage of having a repertoire of specific antibodies to fight an infection that is already partially encoded in their genes. While the immune systems of the indigenous populations could respond, the time between infection and the time required to mount an effective response to a totally new pathogen was longer than the expected survival time of the host. Having never encountered the disease before, they ran out of time and died before their immune systems could fire up.

There are many other infectious diseases that have moulded our genome. Malaria is an interesting case as it has forced changes in at least fourteen different human genes. The ability to fight malaria varies between populations. The malarial parasites produce a barrage of powerful antigens that may overwhelm the immune system's antibody response. The immune system therefore needs to rely on a number of different antibody genes in order to be able to fight back. It is also well known, that in Africa there is a strong correlation between having a natural resistance to malaria, and having the inherited condition sickle cell anaemia. If you have sickle cell anaemia, then you are likely to have a natural resistance to malaria. This is like having to genetically choose between the lesser of two evils.

Studying the inheritance of thalassemia, has also provided us with some useful information about human migratory history. Thalassemia results in faulty protein production for red blood cells and may cause anaemia. There are several different types of thalassemia, each with a different mutation pattern. Through studying one particular thalassemia mutation among different populations in Asia we have found that settlers from islands scattered throughout the Pacific region all originated from the islands around Vanuatu.

But studying the genetic sequences imprinted in the genome doesn't just tell us where our ancestors lived: it can also tell us what happened to them. We update antibody DNA when our environment challenges us to do so. In a period of environmental stasis, there is no need to alter our genome. However, when we change our environment, we have a whole lot of immune system genes and genetic machinery on standby ready to help meet the challenge. It is estimated that there is a potential repertoire of around one billion different possible antibody molecules in a single individual. Yet, so far we can only identify around thirty thousand different human diseases. Less than two hundred are studied widely because all of the others occur at such a low frequency in the population. Of the two hundred that are well characterised, most are congenital, and they involve many genes. These numbers suggest that there is much redundancy in antigen recognition. That is, we may have several different antibody genes to respond to a number of different strains of the common cold.

We now have such a large repertoire of antibody genes that further evolution of the immune system may not be necessary. That is, our immune system is now so advanced that it can respond to almost any new form of foreign pathogen that we might encounter.

RECOGNITION AND RESPONSE

To be able to mount an effective response to any new virus, our bodies must first be able to recognise it as foreign. To assist with the process of recognition we each have an army of T cells patrolling our body.

The T cells are highly specialised white blood cells that have matured in the thymus gland, hence the name 'T cell'. They are the main patrollers of our body. Flowing continuously through our body in the blood, they recognise virus infected cells. The T cells are also very flexible and can penetrate blood vessel walls in their pursuit. Each T cell looks something like a burr, with around twenty thousand identical receptors on its surface.[66] There are between one million and ten million different types of receptors. Yet each T cell usually only has one type. As a rule, each type of receptor is only able to recognize one type of virus-infected cell. This is why we need such a large repertoire of T cells to maintain a healthy immune system.

If a T cell happens to come into contact with a virus-infected cell it recognises, it doesn't destroy it straight away. Instead, it triggers a set of reactions to clone itself. Our T cell count shoots up as an army of ten thousand to twenty thousand identical T cells is produced. This clone army then goes on a seek-and-destroy mission to find and eliminate any other virus-infected cells detected throughout our body. Once all of the virus-infected cells are eliminated, the viral form and the receptors needed to successfully eliminate it are set aside as memory cells.

When the viral invader is no longer detected, the T cells become resting T cells and they resume their normal patrol duties. At this point our T cell count is lowered.

Another important role performed by T cell receptors is to be able to distinguish between the different types of viruses and the antigen DNA and RNA of normal body cells (or 'self'). This is called 'non-self discrimination'. During our early development, the T cells that have receptors that could potentially bind to self are eliminated by the thymus. This is a normal development phase for our immune system. But unfortunately, as we age the range of T cell receptors available to fight new viruses is altered as we continue to fight new viral infections. This means we sometimes end up with T cells that attack our own body. The result is ailments like multiple sclerosis or rheumatoid arthritis that can occur at any stage in our life. Sometimes the onset of autoimmune disease can be linked to a previous viral infection.

CREATING NEW IMMUNE SYSTEM GENES

This is where our other white cells, B cells play a lead role. Like the T cells, our B cells are bone-marrow derived white cells. However, they differ from the T cells in that their surface membrane expresses antibody molecules to fight an infection. That is, they express, excrete and export specific antibodies into body fluids. Each antibody type produced is expressed from one of a number of antibody genes.

When B cells located in special sites in our lymph glands detect a foreign pathogen, some sections of the antibody genes undergo a rapid process of *hypermutation*. The rapid rate of mutation on specific sections of the antibody genes is designed to create a whole new range of B cells in the hope that one of them will be more effective at fighting the foreign pathogen detected.

This means that hypermutation is usually only triggered in B cells that have come into contact with a foreign pathogen. It only results in rapid mutations in those antibody genes that have been able to lock onto the foreign pathogen and destroy it. Immunologists call this affinity maturation. So, the response is dependent upon a specific input from the environment, and it is time dependent. The hypermutation response is also location specific, as only carefully targeted variable regions of the antibody genes undergo hypermutation.[67] Understanding the logic of this process is an important key to understanding just how the immune system has evolved such a large repertoire of new antibody genes.

Hypermutation involves two main phases: one involves direct DNA sequence alterations. The second is an indirect process resulting in a range of new RNA sequences that may become integrated into the DNA. The hypermutation process results in mutation rates that are more than one million times faster than any other known mutation process. The error prone steps involved make it possible to create a massive number of new RNA sequences for the production of highly specific new B cells. However, the mutation rate is so high that more than ninety percent of the resulting new B cells die, mostly due to the creation of aberrant protein structures. The new B cells produced are then sent out to compete with the other types of B cells. Some of these are also new. Here it becomes a battleground based on natural selection. The only ones that survive are those that can bind to an antigen. If the ability for any new B cell to effectively engage the invading antigen by locking onto the three-dimensional shape of the invader is better than other B cells, it will be rapidly cloned. The new B cells are sent out on patrol to fight the infection that they have been selected and cloned to attack.

This process continues until another even more effective B cell is created using the same hypermutation process, or until the infection has been completely overcome. This means that if the infectious agent also mutates, the immune system can continue to relentlessly pursue its role of finding a new effective antibody gene sequence.

Once an antibody that is effective at fighting the viral infection is found, and the infection dies down, the effective antibodies remain throughout the body. They are included as an addition to the repertoire of antibodies already in the immune system. That is, they become a new part of the immune system memory. Here they wait in readiness to fight the next infection. Some new antibodies can be detected in our systems for more than forty or fifty years after an infection has occurred. This is why your doctor can tell if you have previously been infected with a particular infection several years after your recovery!

While the molecular details of the hypermutation mechanism described here are not yet fully understood, it is clear that nature has developed an ingenious way to rapidly create new B cells, and only those that are successful in fighting an invading pathogen survive. This strategy is used by the immune system to enable the body to rapidly mount an effective immune response to almost any new pathogen. Later, they can become reverse transcribed into our DNA for the benefit of future generations.

One example of the immune system in action is when you become infected with a new strain of flu virus. If you have been vaccinated, you may have been injected with a milder but closely related strain of the new virus so that your body is better prepared to fight. It gives your immune system a head start without having to give you the more virulent strain itself. If you haven't been vaccinated for that particular strain, your own immune system needs to quickly go through each of the two steps described to create new antibodies that can effectively fight the virus.

The mechanism of discovery of the most effective B cell is Darwinian in essence as there is no clear evidence that the newly acquired antibody has resulted from any other mechanism than trial and error. Instead of having to generate millions of different individuals with slightly different antibody genes we generate millions of different T cells in one individual. So the discovery of the right antibody is speeded up billions of times and passed onto the germ cells—without necessitating the wasteful process of eliminating the unfit individuals. The processes involved are also Lamarckian in that it is the complex three-dimensional shape of external pathogens that has stimulated the production of highly specific new antibody genes. The updated gene sequence forms a memory of the new foreign invader, and it may become embedded into the genome: both processes are necessary for the immune system to effectively evolve.

ACQUIRED INHERITANCE IN KOALAS INFECTED WITH AN AIDS-LIKE RETROVIRUS

The native koala population of Australia is presenting scientists with an opportunity to study the real-time molecular evolutionary events that occur when a retrovirus invades a mammalian genome and becomes endemic within the population. That is, new DNA becomes integrated into the DNA of a host, and is then inherited by the next generation.

In recent years, the koala population has been infected by an AIDS-like retrovirus called, unsurprisingly,

Koala retrovirus (KoRV). It is having a huge impact on the survival prospects of the species across Australia. The Koala retrovirus is initially transmitted by infection from one koala to another. It then becomes embedded into the DNA of its host. The next generation are born with the viral sequence already in their DNA.[68] By this stage it is said to have become an 'endogenous' retrovirus.

The human AIDS (HIV) retrovirus is an example of a recent new retrovirus to invade humans. It is widely accepted that the sperm and other bodily fluids provide a perfect vector for HIV/AIDS infections to be passed from one person to another. We know that HIV/AIDS leaves a clear signature in our somatic DNA. However, unlike the Koala retrovirus, we do not have any evidence that HIV/AIDS is transmitted to the germline. I am also not aware of any major studies focusing on the intergenerational effects of HIV/AIDS in humans.

The consequences of HIV/AIDS have been devastating. The World Health Organisation (WHO) estimates that the worldwide epidemic of HIV/AIDS resulted in more than twenty five million reported deaths from 1981 to the end of 2007. It invaded the human population by invading the T cells. Without enough T cells, we are unable to fight infections. When the HIV RNA virus enters the T cell, it uses the cell's molecular machinery to produce thousands of copies of itself. It also reverse transcribes its own RNA into the cell's nuclear DNA. From the nucleus it is able to direct and control cellular processes so that it can do its own job in the body.

In the T cell the AIDS virus lies, waiting and ready for the body's immune system to become activated. When this occurs, it instructs the cellular machinery to produce more HIV viral components, and export these out of the nucleus to infect other cells. In other words, it uses the body's sophisticated defence system against itself. Producing the new HIV viral particles causes the host T cells to become damaged and they die. Over time, the number of T cells available in the body to fight infection declines. When there are not enough T cells to effectively fight an infection, the person is said to have fully blown AIDS.

Retroviruses like HIV/AIDS are a common ancestral feature of all mammals. Every one of us has a lot of endogenous retroviruses in our system, but thankfully most are harmless. Many of those that are harmful can be inactivated via successive mutations and deletions. They remain as toothless tigers, providing an historical record of past retroviral infections.

Perhaps the next generation of HIV/AIDS researchers will investigate the intergenerational effects. However, this raises another unsolved dilemma. We simply do not know how new DNA or RNA sequences are so readily transferred between similar or different cell types within an organism. We know that all cells are designed to readily shed and take up infectious and endogenous retroviral material across the cell membrane. Are there special shuttle RNAs to regulate the movement of retroviral material across a cell membrane? These, and other questions raised by the new genetics will no doubt be the focus of research efforts in the decades ahead.

THE IMMUNE SYSTEM IS NOT BLIND

So, it is now evident that many thousands of bacteria and viruses have played an important role in the evolution of our immune system. They have shaped our genome to a greater extent than we previously thought.

In developing our knowledge of how the immune system functions, we are at a stage where we are beginning to understand the important roles played by hypermutation to generate new antibodies and successful antibody selection. While there remains much conjecture about the mechanisms and pathways involved, the result is that the immune system is able to create a new antibody, and then shuttle the new RNA sequence that created it back into the DNA. This requires an environmentally produced infection to stimulate a response, feedback from the immune system via the RNAs, mounting a response, and then the process of reverse transcription to copy the selected successful antibody mRNA templates into the DNA.

To ensure their own survival, pathogens also have a remarkable inbuilt ability to mutate to evade the immune defences of the host, or to increase their resistance to new drugs such as antibiotics. At the same time, it is detrimental to a particular strain if they become too effective at killing their host. The ongoing battle between our immune system and foreign pathogens means that environmental forces are fashioning the human genome, as well as the pathogens we host. Foreign pathogens and humans seem destined to continue to co-evolve, and both will need to continue to develop new genetic strategies to survive. This then becomes an inherently parallel form of Lamarckian evolution.

At the macro-evolutionary level, the immune response is a highly refined and focused set of carefully sequenced actions to defend an organism against a single previously unknown foreign pathogen invading our body from the environment. The response is not random. It is time specific, it requires an environmental trigger, and the resulting changes to the antibody gene sequence are location specific.

It is important to remember that only some of the selected RNA sequence becomes integrated into the DNA as part of the updating process. The resulting changes involve mutations to just a few specific variable regions of the antibody genes, while sequences in other sections remain unchanged (sometimes called 'constant' regions).

Although the immune system is highly sophisticated in how it goes about protecting us from foreign pathogens, it provides us with another example of how our genome is shaped by our experiences as we interact with the environment. The immune system is not entirely blind to environmental stimuli: in the context of the new genetics, it seems reasonable to acknowledge that the evolution of our antibody genes relies on both Darwinian and Lamarckian processes.

* * *

6. EPIGENETICS: THE FOOTPRINTS ON OUR DNA

Just for a moment, imagine that you are sitting on the grass alongside a mountain stream where life abounds. You enjoy the warmth of the sun's rays filtering through the branches of a tree above. A small blue wren carrying a twig from the ground darts up into the air above you, and then into a nearby bush. You feel the velvety damp grass beneath your palms as you turn around to watch. The ground still smells fresh from the morning dew. You feel a small ant hurriedly running along your thumb. In the background you can hear water burbling over pebbles in the stream. A small fish splashes near a branch where a spider is busily weaving her web.

Suddenly you spot a golden marbled crayfish in the soft oozy mud below. In the dim light of the muddy waters you then see a blue one. One is dappled charcoal and white. You see the vibrant red and hot pink claw of another emerge from under a rock.

As you gaze, you wonder just how a few chance mutations in DNA can give rise to such harmonious adaptation, yet spectacular diversity.

While some scientists continue to argue that the forces of natural selection and random DNA mutations could give rise to such splendour, the science of 'epigenetics' is providing us with some new explanations. Over the last two decades epigenetics is helping us to realize how such wondrous diversity and adaptation are generated in all living forms. In the process we are also discovering some of the most convincing—and most curious—examples of acquired inheritance effects.

But what is 'epigenetics'?

The word 'epigenetics' was coined by Conrad Waddington in 1942. Waddington used it to help explain his theory of how the experiences in the life of an organism may cause some genes to behave differently. That is, he believed that life's experiences can alter the way that genes express themselves. As little was known about the structure and role of DNA in the 1940s, this was a rather radical supposition linking our experiences to how our genes behave.

Since Waddington introduced the idea of epigenetics we have discovered why the variations in the marbled crayfish arise: it involves an evolutionary mechanism that is as subtle and as beautiful as the entangled web of life around the stream. We have discovered that many organisms, including humans, respond to environmental conditions by altering some of the chemical markings on the surface of DNA. The chemical markings act as an additional layer of genomic information to alter the way a gene functions. These markings are the 'epigenetic footprints' that literally sit above ('epi-') the DNA. It is as if Mother Nature intentionally leaves her footprints on our genes as we interact with the environment.

The term epigenetics is now used to describe changes in the way our genes behave that occurs without a change in the DNA sequence. We have discovered a variety of methods that our cells have evolved to add the additional information to our genes without altering the DNA sequence code.[69] Some epigenetic changes involve chemically altering molecules on the surface of the nucleotides A, G, C or T. Some involve altering the DNA folding patterns. The process of altering epigenetic markings can happen rapidly. A number of the epigenetic footprints are readily erasable, some are changeable, while others are known to be faithfully copied from cell to cell, and from one generation to the next. Epigenetic processes like these are essential to the development of complex multi-cellular organisms like humans, with each new cell type acquiring a unique epigenetic pattern as we develop.

In recent years the scientific community continues to buzz with new reports of epigenetic acquired inheritance effects. It is now widely accepted that at least some of the new epigenetic patterns acquired in our lifetime can be inherited by our children. Most of the modifications are transmitted with equal efficiency by male and female parents, and a few of the changes can persist for generations.[70] We are also learning that some of our epigenetic footprints can be dynamically altered as we interact with the environment.

These discoveries have helped us to understand why some genes behave differently, and why some organisms born with the same genes can look and behave quite differently. They have also fundamentally altered our understanding of acquired inheritance phenomena.

THE SAME DNA—BUT OH SO DIFFERENT!

The marbled crayfish provides one of the most wonderful examples of how epigenetic systems can be used to produce spectacular diversity of physical characteristics from the same genes. What makes these small creatures even more amazing is that hundreds can arise from a single female who reproduces asexually. While she may have laid four to five hundred eggs, each hatched with the same DNA, there is no crayfish Valentine's Day. All of her offspring live symbiotically with the other life forms in the stream, yet they each look so different. The hot pink ones are hatched alongside the mottled grey and white ones. Some spectacular gold ones with much larger claws may sit amongst them.

But that's not the half of it. This remarkable creature can also alter its own epigenetic system to produce wonderful variations in reproductive behaviour, the number of sense organs, the form of appendages, and its lifespan: but not by its own volition.

Experiments led by Gunter Vogt in Greifswald, Germany,[71] found that physical variations among the marbled crayfish arose in all key developmental stages. In his experiments, it was found that the epigenetic changes are not driven by changes in the environment. Rather, they act as a self-reinforcing feedback loop for the crayfish to create their own diversity within a stable population, even though every member of that population is a clone, created from the same crayfish. This strategy greatly increases the chances of survival of a species when environmental conditions fluctuate. The diversity produced within a population then becomes an evolutionary factor when the environment changes over time.

The main implication is that one single-celled embryo can multiply to produce a wide range of different cell types. In scientific terms, one genome can result in many epigenomes, and thus providing another important set of mechanisms for generating diversity within a population.

Studies of cloned animals have also shown that epigenetic changes may explain some of the phenotypic differences arising in parents and offspring who are genetically identical. In 1996, the first mammal cloned from differentiated body cells was Dolly the sheep. Dolly was cloned from cells taken from the mammary gland of her mother, and named after the singer Dolly Parton. Dolly was the only cloned embryo to survive some 277 attempts. However, she had to be euthanized at age six due to breathing difficulties. Examples of the aberrant growth patterns found in other cloned animals such as pigs, cows and mice include muscular deformation, oversized or undersized birth weight, disproportionate organ development, metabolic disorders and an increase in spontaneous abortions.[72] Some believe that the first generation of cloning technologies failed to transplant adequate epigenetic reprogramming factors contained in the RNAs and other molecular machinery normally accompanying sperm cells to the clone nucleus. We now think that this caused at least some of the observed increase in aberrant growth patterns, and the high mortality rate reported in these early experiments.

While cloning techniques have improved, we are still learning about the importance of other nongenetic factors in inheritance. In the near future, we may be able to readily clone healthy animals. Imagine the pandemonium in the racing world if it was discovered that three different coloured clones of Australia's legendry race horse Phar Lap entered the 2016 Melbourne Cup!

This possibility raises some new questions for humans too. Identical twins who share the same DNA are exactly alike – or are they? Researchers have revealed that epigenetic differences arise in identical twins as they age.[73] Mario Fraga and his colleagues working at the Spanish National Cancer Centre have attracted a lot of attention by showing that young identical twins share common genes and are epigenetically indistinguishable. But by studying a group of thirty-five identical twin sets, they found that as the twins age, the epigenetic patterns vary considerably. Which twin likes to wear bright colours? Which twin has the lighter hair colour? Why is one so sensitive and writes so neatly, while the other doesn't care and takes too many risks in the playground? Experiments involving sets of identical twins have also shown that the variance was greatest for older identical twins who had spent less time together. These twins showed remarkable differences in the overall content and genomic distribution of epigenetic markers that affect which genes are expressed and which aren't.

Experiments like these have helped us to understand why some genes behave differently, and why some organisms born with the same genes can look and behave quite differently. The findings also emphasise the role the environment plays in determining our epigenetic makeup. Although the next generation of experiments may be able to tell us what the differences in specific epigenetic patterns mean, we are learning that we are all capable of dynamically writing—and sometimes re-writing—out epigenetic code. Perhaps in future these will reveal the level of our mathematical ability, how well we are coordinated, what our addictions are, which foods we like, or how often we smile?

The ability to understand how epigenetic markers are linked to our behaviour also foreshadows a new era in forensics. It means that if one of a pair of identical twins—or cloned quadruplets—committed a crime, DNA testing alone will not enable investigators to tell which one is guilty. Witnesses or video footage probably won't be of much use either. Only forensic science based on epigenetic differences would be able to differentiate between the individuals who share the same DNA.

For the present, we know of many epigenetic effects that appear to be a part of a general mechanism for creating diversity among individuals with the same DNA, and that the epigenetic variations are produced at many points along our genes. They appear to be used in many cells and species to create diversity even when the genome and the environment remain stable for a population.

But what specific epigenetic variations do we know of that can arise in our lifetime, and be inherited by our children?

YOU ARE WHAT YOUR PARENTS EAT—SOMETIMES!

What your parents ate could affect the way you look!

Some of the best known experiments on heritable epigenetic change induced by diet, are those that have resulted in a change in the coat colour of mice. What the mother ate, can determine whether her pups are have a yellow coat, a brown coat or a mottled yellow and brown coat.

By feeding pregnant female mice with a methyl-supplemented diet, it was found that the coat colour of offspring can be altered by inducing an epigenetic change in the DNA folding pattern.[74] Researchers have found that the gestating females fed with a methyl-supplemented diet not only altered the gene profile affecting the coat colour of offspring, but it also affected genes relating to the body weight and health outcomes of offspring. Pregnant female mice fed on a methyl-supplemented diet had fewer offspring that were of normal weight, and the health of some deteriorated due to weight related diseases in comparison to offspring from females fed on a normal diet.

The diet-induced changes in coat colour were found to be stably inherited through altered male and female offspring, suggesting the hereditary nature of the epigenetic modification induced by a change in diet alone.

Although there are some other studies suggesting that heritable changes can be induced by a change in diet, this study is important in that it has been able to demonstrate that the change is linked to epigenetic changes on a single gene. Of particular interest is that the mature sperm of affected male progeny show the same epigenetic marks as those created by altering the diet of their mother during gestation. That is, it provides compelling evidence of newly acquired heritable information resulting from how we interact with the environment being transmitted to future generations.

The effect of royal jelly on developing honeybee larvae also provides another wonderful example showing the significance of diet in determining epigenetic changes.

In the case of the female honeybee, it is epigenetic markers that enable genetically identical larvae to become either a fertile queen bee with fully developed ovaries, or a sterile worker destined to go off into the field each day to gather pollen. During the early development stages of larvae, young nurse bees select just a few larvae and feed them the royal jelly that they make so that they will become queen bees. We now know that this is because the royal jelly causes epigenetic changes on a section of the DNA responsible for storing epigenetic information.[75] Timing is important too. The strongest silencing (or switching 'off') of some genes occurs forty eight to fifty hours after injection, which in the real world is a critical differentiation period in the larvae's development.

So it seems that in the case of the honeybee, what is eaten can fundamentally alter their destiny. It alters their developmental fate, morphology, physiology, behavioural repertoires, status in the hive and their lifespan. These epigenetic changes are crucial to the survival of the bee social organisation and the division of labour within it. The changed destiny of the sterile female worker bee needs to be able to rely on its tiny brain to elicit a range of sophisticated behaviours that will enable it to fly, navigate, identify and gather food, mate, feed and fit into the highly complex social organization of a bee colony. The specific combination of expressed genes, and the network of regulatory RNAs responsible for coordinating the signal responses of the expressed genes, suggests that a high level of coordination is required.

There are also several studies on humans showing that what your parents ate can have a significant impact on your future health.

Within the Taiwanese population, it has long been known that chewing the popular betel nut can lead to diabetes in offspring. The more betel nut consumed, the higher the risk. Chewing betel nut leads to glucose intolerance, or the risk of developing diabetes. Diabetes occurs in offspring, even if they do not chew betel nut. The effect has long been known to occur in mice born to affected fathers.[76] In similar studies of humans using parent-child studies, it was shown that children who did

not chew betel nut but were born to fathers who had
chewed betel nut were far more likely to develop early-
onset diabetes. The effect was greater for children of
fathers who chewed betel nut more frequently, and for
a longer period of time prior to conception.[77]

But honeybees, mice and men are very different,
right? When it comes to the way their genes work they
are not that different. Whether you are a mouse, a bee or
you, what your mother or father ate may affect how you
look, and your destiny in a twenty first century world!

MORE MICE TAILS

One of the benefits of inheriting two copies of each
gene is that if one copy is faulty, then there is a good
chance that the matching copy will take over the role.
This is like having a backup set of genes that the body
calls upon when damage is caused to one. So it was no
surprise to discover that there is another epigenetic
mechanism to allow some of our genes to operate
cooperatively. In some instances, only one gene of a pair
will be expressed—but only when given an instruction
by the other.

The instructions given by one gene to another gene
in one generation are also remembered by the next
generation. This is like having a coach on the sidelines of
a soccer field deciding which of his two goalkeepers will
go onto the field next—and how to play the game. When
the next generation of goalkeepers arrive, the coach is still
there to coordinate and instruct the new goalkeepers.

The mechanism that allows genes to cooperate with
each other has been known to occur in plants before
epigenetic phenomena became well understood. As
early as 1956, R. Alexander Brink observed that the

expression of the purple pigmentation genes on the R gene in maize kernels was somehow dependent upon what the R gene was exposed to in the previous generation.[78] That is, they violated Mendel's rules for genetic inheritance. Brink also showed that the effect could persist for many generations.

The same type of epigenetic effect that allows genes to cooperate is now known to occur in mammals. In mice, cooperation between the two genes called Kit has been studied. The Kit gene comes in two forms: in the mutant form, the offspring have a white spot on the tip on their tail and white feet. In the normal gene, these traits are absent. If Mendel's rules of genetic inheritance applied, the combination of these genes inherited could be used to predict whether a mouse has a white spotted tail or not. But this is not the case.

In experiments conducted by geneticist Minoo Rassoulzadegan and her colleagues at the Laboratoire de Genetique du Development Normal et Pathologique in France, the researchers found that some mice born with two normal versions of the Kit gene had a white spot on the tip of their tail.[79] They were surprised to find this.

After further investigation, it was suggested that the information about how the Kit genes were expressed had to have come from the RNAs. They reasoned that the RNA responsible for producing the epigenetic markers needed to produce a mutant form of the Kit gene was the additional factor inherited by the mice. That is, the RNA inherited had somehow provided the information that is crucial for creating the epigenetic marks to alter the expression of a normal gene to become a mutant gene. When a normal Kit gene is silenced by an epigenetic marker, it is expressed as a mutant Kit gene that this is associated with the white tail tip seen in the mice. It seems that the soccer Club Manager had also entered the playing field to pass on a new set of secret instructions to the coach. The coach uses this additional information to alter how he manages the two goalkeepers, and how they play the game!

To test their ideas, Rassoulzadegan and her colleagues decided to obtain some pure RNA from the sperm of mice with white spotted tails, and microinject it directly into the fertilised mouse egg taken from pure bred normal mice that had never been exposed to mutant Kit gene effects. The mice born after the injection showed the white tail tips. Further experiments involving random breeding from the affected progeny showed that the next generation also inherited the same coat defects.

These experiments are among the most important epigenetic inheritance experiments conducted so far: they show for the first time that RNA is an important hereditary factor, and that it is a key factor determining the nature of the epigenetic markers required to alter how our genes are expressed. These experiments are also enabling us to begin to understand how epigenetic effects arise, and how they are passed on from one generation to the next.

Although, much more work is required in this area, what we look like is determined by our parent's genes consisting of DNA, and the RNAs. This raises a lot of new questions about what information we actually inherit from our parents.

EPIGENETIC INHERITANCE OF DISEASE

Just as genetic mutations can give rise to disease, epigenetic errors can affect gene activity and give rise to disease. We know of several diseases linked to epigenetic errors, and some of these are known to be transferred to offspring in rodents and humans. There are numerous studies reporting widespread epigenetic changes in tumour tissue, and some are implicated in the intergenerational transmission of disease in rodents and humans.[80] Two examples are colorectal cancer and melanoma.

Colorectal cancer is one of a number of different types of cancer that is linked specifically to epigenetic changes, although some aspects of this disease remain controversial.[81] There are a few studies reporting that the epigenetic markers identified as being associated with the disease can be inherited, and that some of these epigenetic marks are reversible. Studies like these provide the most convincing evidence to date for acquired inheritance of disease linked to epigenetic errors.[82]

Melanoma is another form of cancer linked to epigenetic changes. Some of our genes that are normally only activated in the early stages of our foetal development are turned on again by a change in the epigenetic markers. Normally, these genes are only activated before our immune system develops.[83] In the case of people who have developed melanoma, these specific genes are switched on again in the cancer cells. We do not know why. While our own immune system can keep the tumour in check for a time, if it is not treated it will eventually spread throughout the body.

Recent evidence linking epigenetic inheritance effects to disease in humans represents a huge shift in our thinking about disease and healthcare. However, the task of tracing the causal links across generations

is daunting. Natural variations in the epigenetic
patterns often show variegated and some quite random
differences within a single population of cells. The link
between the epigenetic patterns and genetic mutations
associated with cancers is also not understood. For now,
the main challenges are to identify those epigenetic
changes that are the primary cause of a particular
cancer, and to understand the intergenerational impact
for humans. While it is likely that in the future we will
probably develop mechanisms to switch target genes on
and off in an adult, the wider implications may not be
known for several decades more. Much of the work on
embryonic stem cells is focused on this potential, and
the hope is that there will be genetic therapies to cure
some diseases related to unwanted epigenetic changes
for the benefit of future generations.

THE MULE AND THE HINNY

All multi-cellular organisms that reproduce sexually,
including humans, are born with two copies of almost
all genes. We each inherit one copy from each parent.
However, a small number of the genes inherited are
'imprinted'. That is, they are silenced, so that in effect
we only inherit one active copy of these genes. The
imprinting mechanism relies on an epigenetic pattern
to instruct one gene to become active while the other
remains in an 'off' state. The active copy is the one used
to generate the functions that the gene encodes.

Around eighty different imprinted genes have been
discovered so far. There is evidence to suggest that these
genes are particularly susceptible to environmental
influences. Some of the imprinted genes are those
that can result in disease like diabetes or metabolic

disorders if tampered with through diet, or behaviour caused by severe emotional upheaval. It seems that what we eat and how we behave in response to outside stimuli can cause disease by changing which copy of a gene is silenced, and which is expressed. Recently it has also become evident that imprinting may play a role in the evolution of a new species.

For thousands of years it has been known that if a female horse breeds with a male donkey, you get a mule with long floppy ears. But if a male horse is crossed with a female donkey you get a hinny that looks rather different, with its strong legs and thick mane of hair. We now know that the differences between the hinny and the mule are caused by genomic imprinting using epigenetic markers. While this phenomenon doesn't provide direct support for Lamarckian inheritance effects, it is providing some remarkable new insights into how our genes function at the boundaries between two closely related but separate species.

Although we do not know the evolutionary purpose for genomic imprinting, we might speculate that one of the advantages it confers is that it can effectively prevent a successful mating between two close species of animal such as the donkey and the horse. In the case of offspring of the mule and the hinny, the result is that a repertoire of epigenetic imprinting effects occurs that is so dysfunctional that offspring often die, or are sterile. The donkey and the horse can continue to genetically diverge as they live side-by-side.

WHY THE CANARY BUILDS A NEST IN SPRINGTIME

While the examples described so far are revealing just how configurable our epigenetic system is, it seems that our epigenetic systems can also involve many genes at once. That is, as we respond to environmental factors, certain environment triggers can result in a genome-wide response involving the simultaneous activation and de-activation of multiple genes. While some can trigger epigenetic changes associated with the progression of certain diseases, others are related to a complex, but related series of behaviours defining how we interact with the environment.

Why does the canary suddenly build a nest in spring? What environmental trigger(s) causes scores of Double Drummer cicadas to suddenly pierce the silence of the summer air after spending seventeen years underground?

It appears that the genome-wide responses triggered are a universal feature of the underlying RNA regulatory networks of all eukaryotic systems. While we can so far only identify a small fraction of the genes linked to the response elicited, we know that a broad range of environmental factors can lead to a global transcription response in which a large proportion of the genome responds at once.[84]

This is like having a whole orchestra in the pit of each cell, ready and waiting for the conductor to raise his baton. When the conductor raises his baton, a single gesture initiates an orchestra-wide response. At first the swirling strings back up the bold brass melodies. Then the sounds of a single violin emerge with the player introducing a powerful and emotionally charged violin fantasy. Later, a single flute rings out over a piano solo. Gradually the whole orchestra joins in and becomes increasingly frantic as it builds to a loud climax, and then dies down. The DNA sequences forming our genes can be compared to the original music score. It is used to guide the conductor and the orchestra. A single subtle signal from the environment may be all that is needed to trigger a genome-wide response that relies on the players and their instruments to all work from the same score, and as controlled by the conductor. The musicians with their instruments are like the RNAs and protein structures that carry out a carefully coordinated set of instructions as required.

But what series of events can lead to such a wholesale activation of a genome-wide response?[85]

Many environmental changes are predictable, such as the differences in light intensity between day and night, or seasonal differences in temperature. The activation of a genome-wide response can also be used to prepare us for the unexpected. In both cases, once there is a genetic sequence linked to these environmental changes, it then becomes possible to add another layer of epigenetic functional complexity to augment existing capabilities. Over time, some responses involving more complex patterns of behaviour have developed that require a coordinated response involving many genes at once.

Nature abounds in examples of just how the complex relationships among our genes and regular changes in the environment have evolved over time. In the case of the canary, their reproductive system is activated when the length of day increases. This triggers complex changes in the canary's endocrine system, and it sets in motion a change in behaviour so that the canary starts to build a nest. If mice are placed into an environment with a low level of oxygen, the growth of vessels is triggered. In both of these examples, changed environmental conditions have triggered a different pattern of gene expression: the logic existing among the genes depends on a pattern of variations in environmental signals, and the result is the coordinated functioning of many links between our genes and our environment. In each case too, highly evolved cyclical epigenetic expression patterns are programmed to prepare organisms for expected environmental change.

Conditioned epigenetic systems also prepare an organism for the unexpected. When yeast cells are removed from a galactose containing medium and placed into a glucose containing medium which they had not encountered before, a global transcriptional reprogramming response is immediately triggered. Once the yeast is 'challenged' in this way, it takes around ten generations before it can grow competitively in a glucose-containing medium.[86] After about ten generations, the reprogrammed adaptations are stably inherited.

In humans, studies have identified two key periods when comparable genome-wide epigenetic reprogramming occurs. Both occur during our early development. It is believed that many of the epigenetic changes established by previous generations are cleared during this process. However, recent research is continuing to uncover more examples of epigenetic changes that escape the reprogramming that occurs when genome-wide epigenetic reprogramming occurs.

* * *

The plasticity of the genomic structures, the RNA regulatory network and epigenetic patterns appears to be a universal feature of all life forms. It enables the genome to rapidly respond and adapt to a wide range of predicted and unforeseen environmental challenges. It enables a wide range of diverse new forms to be created. It also means that we need to alter our idea about what we actually inherit from our parents. Yet we still do not understand the underlying mechanisms. One idea is that the genome might contain a 'DNA library' of possibilities to work with, thus providing the RNA regulatory networks with a set of tools, in much the same way that the immune system genes and some central nervous system genes have developed. But this idea immediately raises another important question. How was such a sophisticated 'library' constructed without environmental feedback? It is difficult to believe that the genes involved were built without environmental feedback.

Looking at the phenomena described, it also implies that the process involves reverse transcription as a general mechanism for updating all genes involved in regulatory and control functions. However it should be noted that the last point remains controversial. While we have evidence that this occurs in the variable region of some genes, we do not have enough evidence to say that it occurs on a much wider scale.

This also means that not only do all living organisms have great inbuilt resilience when challenged by their environment, but they also have the molecular machinery to exploit these challenges to create new phenotypic variation as a part of their natural evolution. From yeast to humans, a range of different epigenetic mechanisms work to actively ensure that we maintain functional fitness in a new environment, or to maintain our functional fitness in a static environment. Like the canary, we all function according to pre-set epigenetic patterns. But we are not at the mercy of all of these. Our genetic and epigenetic systems are designed to enable us to change rapidly or more slowly—depending on how much stress we are exposed to. Some of these are designed to be inherited by future generations.

* * *

7. MEMORIES IN MOLECULES

Francis Crick was among the first to suggest that epigenetic mechanisms might play a crucial role in memory and cognition. Writing in the journal *Nature* in 1984, he stated,

> "...memory might be coded in alterations to particular stretches of chromosomal DNA."[87]

We are now finding that dynamic epigenetic markers are indeed crucial for developing complex behaviours, our memory and our ability to learn. The epigenetic markers on the genome allow the nervous system cells to not only respond and adapt, but to also keep a cellular memory of their previous activity. That is, a molecular recording of many events in everyday lives is dynamically etched as an epigenetic pattern onto the surface of our genes. This is like having our own inbuilt molecular video camera actively recording our experiences complete with sound, smells, images, and feelings!

It has been known for a while that the formation of our memories involves transcription (DNA to RNA) regulation. Yet we have a very poor understanding of how this occurs. Recent research has shown that epigenetic mechanisms are involved in the transcription regulation processes. That is, we are starting to understand the information links between our DNA, the RNAs and other molecular machinery involved in creating epigenetic footprints. While we are busy keeping a digital record of our lives with cameras and mobile phones, it seems that our body is keeping a molecular record of our activities. Some of what we learn is inherited. So now we know that as we pass on the family photos, we also pass on some molecular memories. At the molecular level, scientists are finding that epigenetic markers associated with the memory of behaviour are stably conserved when cells reproduce, and that some of them become a part of the trans-generational inheritance of behavioural characteristics.

Yet it wasn't until 2007 that we discovered that epigenetic changes in the brain are a crucial physical link between our experiences and the formation of memories. To investigate the possible links, Courtney Miller and David Sweatt of the Evelyn F. McKnight Brain Institute at the University of Alabama in the United States of America conducted a series of tests using fear conditioning experiments with rats.[88] Their experiments showed that our nervous system co-opts epigenetic mechanisms to build our long-term behavioural memories.

To build up fear in the test rats, they were placed into a chamber for seven minutes. After the second, fourth and sixth minute, they received an electric shock (1 second, 0.5 mA) on their feet. The fear conditioning protocol induced detectable epigenetic changes that were triggered by the rats' perceptions of their environment. Miller and Sweatt discovered that the induced epigenetic alterations influenced future behaviour, and that a single test trial is sufficient to form a long lasting fear memory. They showed that the most common form of epigenetic change (DNA methylation) is dynamically regulated in the adult mammal nervous system, and that these mechanisms are an important link in the formation of our memories. They also showed that the epigenetic changes involved are not necessarily permanent, but rather they can be reversible.

Through experiments like these, we are discovering the dynamic nature of the molecular mechanisms that enable new behaviours and learning to be passed from one generation to the next.[89] It seems that the molecular basis for our memories provides the means to enhance our existing memories and build new ones through association. Some memories can also be suppressed. This is like having an inbuilt molecular editorial system so that we can later embellish, edit or erase certain memories. Sometimes we simply ensure that we can't access the more painful memories, even though they may remain with us. Memories associated with severe trauma, such as a car accident, are sometimes suppressed.

While this work is helping us to explain how molecular memories are stored ready for recall, are some of our memories inherited?

HOW GOOD WAS YOUR CHILDHOOD?

> Gene_____: *"Depression slowly sets in yet again. I guess its a deadly combo of parents, friends, girls, money, responsibilities, and problems within".*
> *(4:49 pm 19 August 2009, from Txt, Twitter.com)*

There is now an expanding body of research showing that epigenetic changes are induced by our early childhood experiences, and that at least some of these are inherited. As a society, we have not even begun to integrate what this means.

At McGill University in Montreal, Moshe Szyf has identified some clear genetic changes in the brains of male suicide victims who were abused as children, compared to those who died suddenly of other causes.[90] The genetic changes observed were not in the DNA, but were epigenetic. The epigenetic changes observed involve the mechanisms for turning particular genes on and off. But research like this raises more questions than it answers.

It seems that changes in brain cells can be detected long before a person considers taking their own life. Scientists haven't yet been able to tease our whether the changes observed are caused by early childhood abuse, environmental stress, or substance abuse. However, it is hoped that one day we will be able to develop a simple blood test to identify those individuals at most risk of committing suicide.

Some new questions that research like this raises are: Can we design a genetic treatment to erase epigenetic differences? How does a history of neglect and abuse continue to perpetuate through generations? Will psychotherapy or epigenetic engineering be able to make a difference?

As you might expect, these questions are of enormous interest to other scientists in various fields. In answer to the last question, Eric Nestler at the University of Texas Southwestern Medical School, has a theory: "Ultimately we believe that a person who gets better from psychotherapy is inducing changes in the brain".[91]

There are now numerous experiments on rats investigating links between behaviour and intergenerational epigenetic effects that now leave little room for doubt about the inheritance of acquired epigenetic behavioural traits.

One example that has received a lot of attention involves the transfer of maternal effects from one generation to the next. In a high profile study led by Ian Weaver of the Douglas Hospital Research Centre in Montreal, it was found that the maternal traits of licking and grooming rat pups, and arched back nursing behaviour by mother rats significantly alters the epigenetic profiles of their pups.[92] In simple terms, the more attentive and nurturing the mother, the more settled the baby. Baby rats whose mothers were more attentive to them were less fearful than those who did not receive the same level of maternal licking.

Some other differences were also observed. Rats whose mothers did not groom and attend to them so much were more nervous and fatter. The brain cells of the low-licking mothers themselves turned out to have fewer dendrites (or branches from each nerve cell in the brain). While we do not yet understand how our daily experiences trigger the birth and growth of dendrites, Weaver and his colleagues showed that the epigenetic changes measured were long lasting. They also showed that the changes were reversible through modification of maternal behavioural patterns.

Timing of the maternal impact on the epigenetic effects reported by Weaver and his colleagues was also found to be important. Further investigation of the results revealed that the main epigenetic differences observed in these studies only occurred during the first week of life. Because similar studies of the same epigenetic effects on other genes have not yet been conducted, it is not known how common such changes are around birth, or later on in life.

The science of epigenetics is also telling us just how much damage the school bully can do. Studies of mice have found that exposure to aggressive behaviour leads to pronounced social withdrawal, and identifiable epigenetic changes associated with depression.[93] In one series of experiments, it was found that if normal test mice are placed into a cage with an aggressive mouse for ten minutes per day over a ten day period, epigenetic changes associated with the changed behaviour were detected. During this brief exposure, all test mice showed signs of stress and subordination that included vocalisation, adopting a submissive posture and flight responses. This created an identifiable epigenetic pattern. When the aggressor was removed, the epigenetic pattern remained stable for several more weeks before changing back to the initial pattern when no aggressor was present.

More remarkably, some of the chronic stress epigenetic markers induced by the aggressor could be reversed by administering antidepressants. This raises the possibility of treating some identifiable epigenetic markers associated with an illness with a separate type of drug treatment. Perhaps we can engineer drugs to change our behaviour by chemically altering some key epigenetic markers. Eventually we may have an epigenetic antidote to reverse the effects of bullying or parental neglect.

Another important epigenetic study linking acquired behavioural changes to heritable epigenetic effects is by Larry Feig and his colleagues working at the Tufts University School of Medicine in Boston.[94] In their experiments, they exposed young mice to two weeks of an enhanced enrichment program, and compared their long term ability to learn, and their memory with young mice who were not exposed to the enrichment program. The enrichment program used in these experiments included providing access to novel toys to play with, and exposing the young mice to elevated levels of social interaction. This is like regularly placing a very young child into a pre-school crèche where they are exposed to a lot of other young children, and a wide range of stimulating toys.

Feig and his colleagues showed that exposing young mice to just two weeks of an enhanced enrichment program significantly improved their long term ability to learn, and their memory. Their experiments also showed that the benefits were inherited by offspring, even if the offspring had not been exposed to the same enhanced enrichment program or high level of social interaction. They were careful to verify that these intergenerational effects occurred before the birth of offspring. To do this, they split the offspring of the 'enriched' mothers. Half went to 'foster mothers'—mice who had not been through the program—and the other half were raised by 'enriched' mothers. No matter which group the baby mice were in, they showed the benefits of a program that had happened to their biological mother before their birth. There was, however, a difference based on which parent had been enriched: only the females passed on the benefits to their offspring.

If these effects occur in humans, it means that exposing young women to educationally stimulating environments will have some long lasting benefits for a population—even if the next generation are not exposed to the same stimulating environment. However, more studies are required to understand just how some of these sex-linked intergenerational epigenetic effects occur.

While a number of related studies are now emerging, the group of experiments described here provide us with some of the most compelling experimental evidence yet that there are heritable epigenetic effects linked to behaviour. They also show that even subtle variations in behaviour, such as whether or not a mother arches her back while nursing her offspring, are heritable. It seems that just as we leave our footprints in the sand on a beach, your mum and dad, and your grandparents may have all left some of their epigenetic footprints on your DNA.

SYNAPTIC PLASTICITY AND LEARNING TO SWIM

When we learn to swim, to play golf or to play the piano, the ability for our neuronal cells to strengthen or weaken their individual connections following neuronal activation gradually builds up a type of molecular memory. Through use, and lots of practice, it helps us to improve our performance over time. Epigenetic mechanisms are now known to play a crucial role in the development of a wide range of learnt physical activities that are influenced by pre-set signalling patterns in our neuronal networks. The process of strengthening or weakening individual connections following neuronal activation is called 'synaptic plasticity'.

In the case of a snake that can move without a need for legs, the opposite effect might occur—demonstrating the effects of disuse of parts on our genomic expression patterns. The use of legs requires the constant release of a series of proteins involved in the coordination of movement. This assumes ready access to a certain set of specific DNA sequences so that transcription and translation can be initiated and coordinated. The DNA folding structure needs to make these sequences readily available for high levels of expression as required by an organism with legs. As the snake does not need leg movement, it is no longer necessary for the genome to make the sequences associated with this movement readily available for expression. Over time, these have probably become suppressed and hidden deep within the condensed DNA, or even been deleted. The set of sequences that need to be expressed for slithering are readily available. We can only guess at how differences in the use and expression of a particular set of genes are related to the embryonic development of a snake that results in its forming only tiny dysfunctional remnants of legs. Discovering more about this might provide a molecular explanation of what Lamarck described as the 'the use and disuse of parts' and Darwin called 'functional shift'.

In the process of improving the signalling systems in our own neuronal networks as we learn to walk, we can reshape our muscles and tendons. Bone remodelling (bone 'plasticity') might also be involved to slightly modify the supporting skeletal structures to further improve our fitness and our form for walking.

Bone remodelling provides another good example showing how genes controlling our central nervous system and our physiology are designed to work cooperatively as we interact with the environment. We all experience considerable changes in our skeletal structure from the time we are born, and until we become frail and old.

Throughout our life, bone continually remodels itself. Experiments using mice by Florent Elefteriou from the Baylor College of Medicine in Houston, Texas and his colleagues,[95] have found that a hormone produced by fat cells called leptin directly regulates bone physiology through control of the sympathetic nervous system. Although we do not yet fully understand all of the mechanisms or pathways involved, the body remodels bone so that heavier women have higher leptin levels and high bone mass, whereas lean women have very low levels of leptin and a low bone mass. It seems that the heavier we are, the more likely we are to have a higher bone mass to support the weight.

A wide range of examples of synaptic plasticity and its link to epigenetic markers are now being studied. These include learning new taste sensations, spatial memory formation, and the establishment of emotional memories such as fear that is conditioned.[96] Other examples include schizophrenia which alters one's sense of reality, cocaine addiction that profoundly alters the brain's reward pathways, alcoholic addiction, depression and a predisposition to stress. Cocaine produces specific epigenetic modifications within thirty minutes of a single hit. Each of these examples highlights the complex contributions of multiple and related epigenetic mechanisms. We have found that all of these examples are associated with specific long lasting epigenetic modifications.

Some related studies have also investigated whether or not the associated epigenetic modifications are inherited. Several studies of normal twins have shown that heritability for schizophrenia is estimated to be greater than eighty percent.[97] That is, if one twin is diagnosed with schizophrenia, then the chance of the other twin also being diagnosed with schizophrenia is greater than eighty percent. In the case of cocaine addiction, researchers investigated whether or not the epigenetic patterns associated with the dispositions can be reversed. It was discovered that the cocaine-induced epigenetic alterations activated can be reversed.[98] However, it will probably be decades before we really begin to get on top of these.

The point is that we are beginning to realise that many different types of human memory and cognition are associated with epigenetic changes. The epigenetic phenomena associated with synaptic plasticity are now becoming quite well understood. We now know that long-term synaptic plasticity involves changes in gene expression showing the importance of epigenetic mechanisms for gene regulation in both vertebrates and invertebrates. The epigenetic effects associated with some dispositions are reversible, and some are known to be inherited.

More importantly, these phenomena provide a molecular understanding of how the use and disuse of parts, human memory and cognition might all alter our evolutionary direction. They might also help to provide a molecular explanation for second generation sporting talent, mathematical ability, or for a fourth generation musical genius.

THE NEXT HOLY GRAIL

The first structural descriptions of a new family of central nervous system genes called 'cadherin' genes were reported in 1995. Although there is still much to learn about this super family of genes, they are starting to shed a new light on the molecular and genetic basis for our understanding of the development of our memories and neuronal structures.

The most surprising feature of this remarkable new family of genes is that there are variations, and they are found in organisms ranging from the unicellular *choanoflagellates* to all classes of vertebrates. Detailed analysis of their molecular evolution provides evidence of their adaptive evolution over billions of years.[99] It seems that our ability to interpret and coordinate a response to the world around us has relied on the development of this amazing super family of genes over billions of years.

The main role of cadherin genes is in cell-cell adhesion and communication. They can act to control the adhesion and communications between like cells, or dissimilar cells in a way that leads to the coordination of multi-cellular structures and even whole organisms. Some autonomic nervous system genes such as those required for cardiac functions rely on cadherin genes to enable the central nervous system to mount and coordinate an automatic response to environmental signals that are 'hardwired' into our DNA. We rely on cadherin regulation to provide the cell-cell communication links required to keep our heart beating away at around sixty to seventy beats per minute. A tap on the shoulder from behind and your heart suddenly pounds! Different pathways may then become dominant in a particular cell type, and these may be influenced by either positive or negative feedback mechanisms using an expansive network of input signals that connect all of our organs and tissues, including the brain.[100] Thanks to the rapid transcription events involving the cadherin genes, you are now ready for action. You look at who it is, and decide how to respond.

What makes this family of genes so interesting to geneticists is that we are beginning to discover that like the immune system gene family, they have some Lamarckian properties. We are now in a position to investigate the nature of the molecular pathways involved in giving them their remarkable Lamarckian properties. An important question to be asked by scientists in the decade ahead is: how is new genetic feedback regulated to limit what becomes integrated into the variable region of the cadherin genes? Further research will no doubt improve our understanding of how this occurs, and the role this has played in the evolution of the human brain.

Over the last decade scientists have discovered that some cadherin genes exhibit some of the same characteristics as the immune system genes in terms of how they function, how they become updated in response to environmental stimuli, and how the genes themselves are structured and organized. Functionally, the human immune system and the central nervous system have a lot in common. They share the capacity to recognise the unknown, and the ability to mount a response to an almost infinite range of possible threats. Both systems use the rapid secretion of molecules such as hormones into the bloodstream so that they can influence events in almost any part of the body. Both systems also rely on an extensive network of nodes and links throughout the body to be able to coordinate a whole body response to a threat. Like antibody genes, some cadherin genes have variable code regions where new DNA sequences can potentially be inserted, as well as highly conserved segments of DNA called constant regions. These undergo similar types of genetic rearrangement, and there is some initial evidence to suggest that the process of hypermutation is also used as one of a number of steps involved in updating the variable region code as the genes respond to environmental signals.[101] There is evidence that several events of extensive sequence deletions and reshuffling have occurred in their evolutionary past. That is, it seems likely that the cadherin gene family has adopted the same evolutionary strategy for updating memory in our genes as that used by the antibody genes that have been intensely studied since the 1950s.

The human immune system and the central nervous system are also quite different in that the immune system genes rely on antigenic stimulation by a foreign pathogen from the environment, while cadherin genes rely on central nervous system stimulation. The main physical difference between them is that the immune system relies on a liquid flow for cellular transport, while the central nervous system relies on a fixed network of neurones and dendrites. The nervous system uses a network of dendrites (nerve cells) that use a chemical signalling system that relies on electrical depolarization effects that zip along the connections extending from the nerve fibres at an extremely fast pace. Recent research has shown that the dendrites are tipped with highly flexible molecular structures ('growth cones') that are attracted or repelled by a large variety of different chemical signals. The variable regions of cadherin genes are capable of generating a large number of different forms, consisting of multiple isoforms linked together like pieces in a jigsaw puzzle. They can package and repackage previous sequences, and sometimes new sequences are added. The re-packaging of existing sequences can be compared to the way we use words and sentences to communicate a message, or to re-use old concepts to create a new message. The process involves firing across a network of neurones when responding to stimuli. The large number of network possibilities gives rise to an almost infinite number of possible new somatic variants of RNA.

It is not yet conclusively established that the RNA variants are reinserted into the DNA using reverse transcription. This is only hinted at so far.[102] Future discoveries will need to establish whether or not these genes undergo reverse transcription as inferred from the data on antibody genes, and how the process is regulated for the different types of cadherin genes. Although when we look at Figure 1 in the Appendix, it seems logical that this would occur. If the information flow involved in reverse transcription (RNA to DNA) is found to be used by more gene families, then it would help us to explain how the highly specific patterns already inserted into our DNA were created. This seems to be a missing link, as we know that highly specific DNA sequences are used to create copies of RNA to trigger 'hard wired' instinctual responses involving the central nervous system genes. Future research will reveal how epigenetic memory (or 'soft' wired memory) can enter the genome using reverse transcription in the variable sections of genes. For the present, this idea remains one of the scientific 'Holy Grails' separating the neo-Darwinian scientists from those that now openly entertain a Lamarckian view of evolution.

Although we still have much to learn about this particular super family of genes, it is becoming increasingly evident that the immune system and the central nervous system genes use some of the same genetic strategies to learn, to remember, and to update our RNA and DNA. RNA editing functions appear to provide an important set of molecular tools for modifying gene function by overwriting the original code. This phenomenon is well documented for encoding proteins involved in fast neural signal transmission in parts of the nervous system network such as the serotonin receptors. This type of RNA editing is known to be far more prevalent in humans than in mice.

The central nervous system genes and the immune system genes are also designed to work cooperatively. Psychosomatic effects linking the nervous system to the immune system have been known for centuries. This is why we get a cold just after we finish an exam and the stress has left us. We know that under acute stress conditions, an infection may be held at bay, but the immune resistance collapses once the stress is removed. The reason is that mounting an effective immune response requires a complex coordination of both systems. Similarly, psychological trauma or illness makes a huge demand on the nervous system.

Experiments verifying that immune system responses can be conditioned provide us with some further evidence that the two systems are closely linked, and that they have co-evolved over long periods of time. In 1975 Ader and Cohn[103] published experimental results that showed that if an immunosuppressive drug was administered with the taste of saccharin, then after a time, the taste of saccharin alone would produce the same immunosuppressive response. While some were sceptical about this work when it was first reported, the results have since been verified by other researchers studying the placebo effect. The immune cells can produce chemicals to transmit signals to other cells, and express the receptors to respond to them. For example, lymphocytes used by the immune system bear some receptors that are used by the central nervous system. Lymphocytes also possess binding sites for opiates, yet the main site for opiate action appears to be in the nervous system.

It is therefore not surprising to discover that both the immune system and the central nervous system express some of the same genes. For example, in the nervous system, the genes associated with neuronal growth and signal transmission are the same as those used for facilitating contact between the immune system's T cells and antigen presenting cells. There is now substantial evidence indicating that there are mechanisms for constant two-way communications between the immune and nervous systems. They also share the use of a large number of protein structures, they share some genes, and they both make use of some of the same RNA regulatory machinery. By using two such coextensive networks, a fast immune response can be quickly activated in a localised area of inflammation when a sudden injury occurs. It also provides us with further insight into the co-evolutionary directions taken by different gene families to ensure that we continue to function as a fully integrated organism.

SHORT TERM AND LONG TERM MEMORY STORAGE

The latest idea emerging is that lifelong memory storage is not limited to the epigenome, but that it involves more permanent changes to the genome as well through the process of reverse transcription.[104] How else did the carefully sequenced genes relating to our behaviour become encoded into our DNA and ready to be actively turned either on or off?

Understanding the molecular mechanisms involved in generating diversity, selecting and cloning useful new sequences, and then integrating new sequences into the DNA for future reference is therefore now receiving a lot of attention by those working on central nervous system genes. Previously, we believed that our heritable information was stored only in the DNA. However, the collective findings in the broad field of epigenetics now imply that our own memory storage operates at two main levels.

The first level of memory storage involves the creation of a semi-permanent record that resides at the level of the epigenome. At this level, epigenetic elements that include a mix of associated RNAs and proteins can be thought of as forming a network of bridges linking our genes with the environment. The formation of epigenetic memory can last anything from a few seconds to several years, and it can be passed on for several generations without any further modifications to the genome. Some epigenetic memories are reversible in a single generation.

The second level involves the creation of more permanent genomic memory. It involves copying or updating a record of some selected memories into our genomic DNA using reverse transcription. Although it has not yet been verified, there is now indirect evidence suggesting that reverse transcription provides the primary mechanism for this to occur, and that the molecular steps involved for central nervous system genes are similar to those predicted to be used by the immune system genes.

Functionally, it is also a requirement that any new genomic information is packaged with some special instructions on how and when to use the new information. There needs to be some additional sets of coded instructions associated with the new information to identify appropriate triggers and rules governing regulation for the expression or silencing of new genes. This could be in the form of DNA copies of some small RNAs that are reverse transcribed as tiny extra passengers into the new genomic sequence. For example, the additional DNA sequence information might include some instructions linking the activity of one gene to another gene in an entirely different gene family.

As a complex information management system, the two layered molecular memory storage system is well coordinated, and highly regulated.[105] It has inbuilt error checking and it is capable of updating itself with new information at all levels. At the level of the epigenome, it is capable of generating extraordinary diversity. The molecular memory elements that epigenetic systems rely on are also incredibly mobile. They are mobile at the level of the whole organism, organs, tissues and the individual cell. In fact, our biology presents a superb architecture for building a highly sophisticated memory system, and it is linked to a self-learning central processing unit (the brain) that is far superior to any artificial intelligence program developed by us.

Another outcome of recent epigenetic research is that the inheritance of new memory appears to be far more complex than previously thought. We are really only beginning to scratch the surface in terms of our scientific understanding. What we are uncovering is a set of molecular tools with layers of complexity, redundancy, and incredible efficiency. They provide us with the ability to rapidly acquire new molecular memories, and then pass them on to future generations. Mapping a one-to-one correspondence between the genetic code structure and phenotypic features is not possible when we look at the mechanisms at play. We need to view each level as a new and more complex set of rules governing the flow of information between the environment, the somatic cells, the epigenome, the universe of RNAs and proteins, our genes— and the next generation.

The science of epigenetics is also demonstrating the importance of the regulatory roles played by the network of RNAs. The main reason that RNA networks are more effective and efficient than proteins is that they are far smaller and highly mobile within cellular structures. They can rapidly mutate or restructure while performing their role. Each small RNA can be viewed as a tiny piece of mobile code that performs a single function. With their mobility and high mutability, the combination of RNA activities and proteins present many opportunities to receive and manage environmental feedback. In this sense, the RNA regulatory network is the Gatekeeper of the genome.

All of this seriously challenges our definition of a gene as a static, passive entity. The new genetics is providing us with a molecular understanding of how environmental feedback can become incorporated into our heritable material. How we react to environmental factors, what we learn, what we do, and what we eat will all potentially have a long term intergenerational impact on our genetic pool.

While most scientists still remain cautious about articulating the Lamarckian implications, the distinct Lamarckian flavour of the new epigenetic world unfolding before us is becoming a strong antidote to genetic determinism. In the meantime, the message emerging is that we each have the ability to erase, write and re-write our own memories in molecules—and then pass some of these on to the next generation.

* * *

Dear Reader,

When I thought of you reading this book I suddenly recoiled with the realisation that the molecular basis for my thoughts as a writer might be dynamically copied straight into your epigenome as you read. While the copying process is not a high fidelity one, given the different beliefs we harbour, the consequences of our synaptic plasticity means that we do not need to physically exchange epigenetic or genetic material to reproduce a copy of our own living code in others. Through books, laughter, language, art, or music, we all leave epigenetic footprints behind.

Carrying these thoughts forward leads to a future in which we will all become aware of the epigenetic footprints we leave behind as we interact with others. This is a new type of environmental challenge.

I paused...and recoiled again!

In Aldous Huxley's novel *Brave New World* that he wrote at the height of the last Great Depression in 1931, the drug 'Soma' was used by the government as a method of controlling people through pleasure. It became a popular dream-inducing drug carried by all citizens to enable them to escape the hassles of everyday life. 'Soma' is also the name given to one of the drugs that Michael Jackson was reported to have taken prior to his death. What role will drugs that manipulate our epigenetic and genetic profile or our molecular memories play in the future?

* * *

8. LAMARCK'S NIGHT OF DOUBT

The Greek philosopher Plato (429-347 B.C.) wrote a series of notes summarising the things he had learnt from his teacher, who was the philosopher Socrates. One of these is a story that has become known as 'Plato's Cave'.

In the cave there were prisoners who were chained so that they could not turn their head. They only ever saw the shadows on the wall cast by the fire behind them, and the echoes from within the cave walls. The prisoners did not recognise their imprisonment as this was the only existence that they knew. Some prisoners escape however, and discover a world with sun, grass, trees, rivers and real people. They go back and tell the others what is outside of the cave. But when they return, they are ridiculed as madmen.

Of course Plato's message is that it is a struggle for us to extend our realities beyond our existing perceptions and beliefs. Plato likened our life to that of the prisoners. We are all prisoners in Plato's cave, and most of us will not understand messages brought into the cave from outside. In the cave of evolutionary thought, even if someone brings a new message, most will choose not to listen.

Maybe at present, the new genetics and the implied Lamarckian messages it brings back to our cave are just a few early warnings that we need to augment our beliefs about evolution. But never doubt that the Lamarckians returning to the neo-Darwinian cavern of evolutionary thought will continue to be ridiculed.

* * *

For most of the last century the neo-Darwinian view was accepted as a scientific fact. Yet, it was evident to some of us that the ideas it embodied were essentially irrational. Two of the more influential arguments that the neo-Darwinians could not explain were based on the idea of punctuated evolution, and the mathematical improbability that our life forms arose by random mutations and selection alone. The original proponents of these ideas did not have the benefit of large amounts of molecular data.

WHY ARE SOME FOSSILS MISSING IN THE EVOLUTIONARY RECORD?

Back in the 1970s, Niles Eldredge and Stephen Jay Gould proposed their idea of punctuated evolution.[106] They noted that the fossil evidence suggests that the process of speciation does not make continuous progress. Some species can remain in the same state for millennia, while others change so rapidly that there is little evidence linking changed forms in the fossil record. New species can arise rapidly during brief periods of rapid change.

Gould argued that only rapid speciation explains why fossils of some intermediate forms that might occur during brief periods of rapid evolution over a few tens of thousands of years are not found. The punctuated evolutionary theory as put forward by Eldredge and Gould, seems to be the rule across species, rather than the exception.

Eldredge and Gould, and other supporters of their theory claim that there have been some periods of such rapid evolution in the past that it is impossible for it to be the result of natural selection alone. Although many evolutionists consider change over millions of years, rapid evolution can occur in as little as fifteen thousand years. Eldredge and Gould argued that this is the amount of time it has taken a large variety of cichlid species of fish to appear in East Africa's Lake Victoria.[107] The Lake Victoria basin was entirely dry around 15,000 years ago, probably due to an extremely dry climatic change. As evolution requires new information to be generated, random mutations and natural selection cannot account for the large number of new genetic sequences required to produce a new species in the time available.

If we view an organism simply as an expressed form resulting from a set of genes that might undergo random mutations, then we are missing the point. Picture having a huge castle built of Lego blocks, and then relying on random changes to the blocks to create a model of a twentieth century high rise apartment block that is perfectly suited to its surrounds. All living organisms consist of a whole interactive network of genetic and epigenetic systems that are intrinsically linked to the environment. As such, they respond to stimuli using feedback mechanisms operating through complex and pre-set pathways. While we do not understand all of the mechanisms for regulating whether or not, and which type of information actually gets fed back down to the genome, we do know that it happens in a highly coordinated way. But the most profound implication for evolutionary theory is that such significant change could lead to an incredibly rapid, and at the same time orderly change in a species and speciation effects.

In this sense, a holistic view based on a Lamarckian view of evolution provides a causative explanation for the punctuated evolutionary phenomena.

We know that it is possible for a massive rearrangement of the genome to occur in environmentally stressful situations. We also know that such responses are built into a threshold system. Rapid increases in temperature, the presence of water, electromagnetic radiation, the amount of oxygen in the air, and in the type of food available are just some of the environmental stimuli triggering changes in the genome and epigenome. If the earth's climate changed suddenly due to a meteor shower or a rapid increase in volcanic activity, then it is possible that such an event could result in a period of rapid evolutionary change. We could expect that an explosion of diverse genomic and epigenomic forms would be generated. Many species may disappear. Generated by direct feedback from the environment, the more extreme changes would put the epigenomic and genomic co-evolutionary machinery into overdrive, and new species could be generated.

The crucial divergence here from a neo-Darwinian view is that such changes are not random or blind. The genomic elements for change already exist for adaptation to a different food supply, a wetter environment, or more oxygen. The changes produced could also co-evolve to generate other improvements in editing and transcription processes to retain overall genomic stability at times of greater environmental stress. These become important when high oxygen levels, or increases in temperature which are known to cause an increase in mutations, occur. The pangenomic molecular machinery is also there to cannibalise existing systems to enable an organism to access and digest different food sources.

Such large scale molecular updates can co-evolve rapidly across several species. The environmental impact, and competition among species will all act to rapidly drive the genomic and epigenomic systems. Rapid reshuffling of genomic and epigenomic elements, hypermutation, reverse transcription events, somatic hypermutation and a host of other molecular gymnastics all contribute to the process of change. During these periods, the nature of the genetic or epigenetic changes is triggered by the environment, and they co-evolve with the environment. The regulatory genes serve to ensure that the rapid process of change is coordinated and regulated. Their role is therefore governed by internal factors, and the ability for extended cooperation across genes, or gene families already exists. It is this that makes the underlying process of genomic change directed, rather than random.

These principles of organisation have been built into genes since the earliest known life forms appeared on earth. Our scientific understanding has developed to a point where we are able to understand this, and to discover that the capacity for adaptation is co-extensive across a larger and more fluid pangenome. A key starting point is the capacity to develop memory in the form of a living library of remembered patterns stored in the genome and the epigenome.

Of course, none of these ideas alone can provide definitive answers about our evolutionary past. All they can do is offer a different framework for re-considering how rapid evolution can occur.

SO MANY MIRACULOUS ACCIDENTS IN DESIGN

Other efforts to discredit the neo-Darwinian view relied on the mathematical improbability that our design arose by chance mutations and natural selection alone. The idea of natural selection acting on random mutations cannot adequately explain the many examples put forward. However, the emerging paradigm based on our ability to rapidly respond and update our genome offers a plausible new framework to reconsider these arguments.

So far we have learnt that all genomes consist of an ever-changing network of pangenomic elements. Humans carry enough nucleic acid to make over three million genes. We pack these into a flexible bundle so that they fit into the nucleus of a cell. Some gene sequences slip along so as to re-order their position. Some cross over, some have their sequence updated, and their surface chemistry can be altered. New genes can be added and in other genes extensions can be added, while some sections are deleted or replicated and checked for errors. Many elements are highly mobile and versatile. Some fine-tune other genes or protein structures. Some control the body plan. Each gene can contribute either directly or indirectly to different structures or functions within the body and cells. And any given function relies on a number of different genes. All genes are interacting with one another, and with the RNA regulatory network. All genes are either directly or indirectly linked to sensors and signals received from the environment. To build a cell, a tissue, an organ or an organism and a social structure to support a whole population of organisms requires a continuous process of diversity and adaptation. It also relies on an underlying co-extensive network based on relationships between all of the interactive elements at each level.

So how did so many miraculous co-incidents in design evolve by chance mutations and natural selection alone over the timeframes available?

The reasoning used by those arguing against a neo-Darwinian view of evolution goes something like this: the mathematical probability that the genomic structure of a human was formed from random mutations and natural selection in the timeframes available, is so small as to be logically impossible to accept. This reasoning is used to erode the neo-Darwinian edifice on the basis of the improbability that life itself arose by chance, and that the diversity and complexity of life forms were generated by random mutations and natural selection alone.

What the new genetics is telling us is that our heritable material has evolved with us using a myriad of molecular mechanisms to reflect how we interact with our environment in each generation. That is, even a single generation can play an important role in determining our evolutionary direction.

What is the probability that random mutations gave rise to the callusing of the breast of an ostrich chick before it is born so that it can rest on the ground like its parents? What sorts of random mutations caused the legs on a snake to become so small and dysfunctional? Just how did the brain of a bee develop by random mutations alone? Our own biology too presents some fascinating examples of ingenious design. What is the probability that random mutations resulted in something as complex as the human brain?

In the context of the new genetics, how could millions of random mutations across our genes simultaneously give rise to coordinated and directed changes in related genes or unrelated genes? How could chance mutations and natural selection alone rapidly produce whole families of new genes, and interrelated genetic sequences?

There is a huge range in the length of our genes. A single simple protein coding gene can be between three hundred and one thousand base pairs long. Some of the more complex genes can be up to one hundred thousand base pairs long. This gives rise to an astonishingly huge range of potential sequences. Yet, DNA synthesis is a spectacularly high fidelity process, with only around one unforced error for every 10^8 bases copied. At each point in the gene in our DNA, an A, a G, a C or a T is found. That means at each point there are at least four possibilities. If we take a tiny gene segment of say two hundred and fifty six base pairs, then there are 1.3×10^{154} different possible sequences. The number of potential sequences for longer segments is huge! How did the primitive mammals that existed one hundred million years ago give rise to a new and highly complex species in this time? While we still use the same universal code, the number, form and length of some genes has changed, and new genes have been added to enhance the genomic repertoire of extant mammals. The evolutionary learning algorithms operating on our pangenome have enabled us to develop more sophisticated immune responses, to develop highly coordinated central nervous system responses, and to introduce a raft of new coordinated behavioural and developmental genes.

So how were an estimated six hundred and eighty nine new genes generated in man since diverging from the apes only six million years ago?[108] Ad agencies and biologists the world over have been telling us for years that we share ninety eight percent of our DNA with chimps. Based on more recent studies of mammalian gene families, it is claimed that the human and chimp genes differ by at least six percent, which is far more than previous studies have concluded. But in the context of the new genetics these numbers don't mean much. As a result of the genomic revolving door resulting in new genes, and gene loss, and epigenetic mechanisms to vary how genes are expressed there is an even greater number of differences between species, and between individuals within a single species. There are a huge number of variations at the level of nucleotides in genes, variations in gene structures, and variations in gene families. If we add the accumulated and lost differences arising from epigenetic effects, then it is possible to generate even greater diversity. It means that we can potentially differentiate between all individuals regardless of their DNA, as well as the cell type and the age of each individual. The gain and loss of so many genes, gene families, and the altered functions of some individual genes from one species to another provides a fertile source of adaptive change that cannot be explained by random mutations and natural selection alone.

In a vivid take on the origin of life itself, the great astrophysicist Fred Hoyle is famously known for describing the possibility of life evolving by chance as being as improbable as a tornado blowing through a junkyard and forming a jumbo jet. As we grow to understand more about the pangenomic differences between each species, the probability that life has evolved since then through competition and random mutations alone becomes even more mathematically absurd: we can continue to believe in a neo-Darwinian world if we believe that one hundred Airbus A380s can simultaneously explode in mid-air into hundreds of thousands of components, and then randomly land on the same runway in a carefully ordered sequence, and in a fully functional form.

Even if chance mutations gave rise to the variation evident today, then the neo-Darwinian view based on random mutations and natural selection is still not able to explain how new genes arise, or how whole new sequences become embedded in the genome without altering other highly conserved sequences along the way. It also doesn't explain how the new gene structures and functions arise, or how new chromosomes came into existence.

This brings us to a point where our fundamental beliefs and the ruling scientific paradigm of the day are being seriously eroded by the quality and quantity of some now irrefutable facts. Many in the scientific community are no longer so inclined to reject the notion of Lamarckian inheritance. As recently harvested genomic data is analysed, scientists are finding that at the very least the neo-Darwinian paradigm needs to be updated.

So how much scientific evidence do we need to move something from the realm of theory to fact?

WHEN DOES INFORMATION BECOME A 'FACT'?

In science, a few raindrops usually just form a puddle—and people can conveniently step around it. But enough raindrops, and eventually you will get a river that flows into a much bigger ocean of truth.

A wide range of scientific information now verifies at least some modes of acquired inheritance effects as a 'scientific fact'. Yet evoking the process of renewal for our scientific thought is not easy. Our collective scientific learning is done by individuals and groups working within the confines of entangled social and political structures. This means that only when the whole scientific community is confronted by a fundamental problem does it find ways of addressing it.

Why, then, in the face of so much sober and sound evidence spanning several disciplines are there scientists and science writers who resist the idea of acquired inheritance effects so vociferously? Our collective scientific mind has to agree where to erode the previous boundaries of what is, and what is not an acceptable scientific idea and worthy of further investigation. Thus it seems that the idea of what is acceptable science regarding our perceptions of how evolution occurs has become a formalized process created by the human mind to justify the high costs of searching for truth.

The discovery that the world was not the centre of the universe and its importance to science, remind us of the tensions between scientific reasoning, politics and faith. In 1543, the Polish astronomer Nicholas Copernicus published his theory that the earth revolved around the sun. The Roman Catholic Church condemned Copernicus for his theory as it removed the earth from the centre of the universe. In 1611, the Copernican view of the universe was confirmed when the Italian astronomer and mathematician Galileo Galilei described his findings on what he discovered through his amazing new brass telescope. According to Galileo's observations, the earth did revolve around the sun. The Roman Catholic Church condemned Galileo for supporting the Copernican view of the universe. They refused to even look through Galileo's new brass telescope. Under the weight of papal authority and to avoid being burned at the stake, he was summoned to Rome and ordered to stop teaching or writing about Copernican theory. He spent the last eight years of his life under house arrest. Even though telescopes were available it took almost four centuries for the Vatican to recognise that the Roman Catholic Church may have been wrong in condemning Galileo for supporting Copernicus.

In 2000, after thirteen years of investigation by the Vatican Science Panel, Pope John Paul II apologised to the world for condemning Galileo and sending him to jail.

In an ideal world, science allows highly specialized individuals like Copernicus and Galileo the freedom to pursue an endless search at the boundaries of our reality. While these individuals limit the view we see at the edge of our world, our political, social and religious beliefs work together to further restrict what we are told is truth in science. The story of Galileo is used as an iconoclast illustrating that real science is both a social construct, and a framework within which individual scientists are confined.

Much has been written about these processes and the relationships involved in permitting the birth of a new scientific paradigm. In the literature on the history and philosophy of science there are many examples and views arguing that science is not about the monolithic body of knowledge we teach in our schools. It is a socially constructed set of rules to further the quest for new knowledge for us as a society. In this sense it reflects our changing social goals. It reflects how we perceive what is around us, and it is built on top of our underlying belief system.

In Thomas Kuhn's work 'The Structure of Scientific Revolutions',[109] he used the term 'paradigm' to describe a basic concept used by scientists and generally considered to be self-evident. Using this concept Kuhn argued that scientific knowledge does not grow linearly with the gradual accumulation of knowledge, but that its progress is marked by revolutions. During the period when a paradigm is strongest, results that do not conform are considered to be an anomaly. Or worse, they are completely rejected. This is what has occurred in the twentieth century when the reign of the neo-Darwinians reached its peak. However, as the tide of new knowledge that refutes the reigning paradigm grows, Kuhn argued that science eventually reaches a crisis point. At this point, a new paradigm may emerge which subsumes the previous one. Kuhn referred to this as a 'scientific revolution'. Only then does it become acceptable for maverick scientists to function within a new framework, when previously they may have been ostracised by the scientific community.

At the peak of a paradigm, the basic concepts it embodies become so invulnerable that they can also become irrational. In a democratic capitalist society based on survival of the fittest, scientists have continued to build their scientific reputations using neo-Darwinian logic. Bolstering each other, our scientific, political and economic principles all acted to re-enforce each other. Phalanxed by several high profile popular science writers, the neo-Darwinians have held centre stage to now. Yet, if we look at the collective works of prominent scientists, popular science writers and many educators of the last few decades it is evident that the idea that evolution arose as a result of random mutations is essentially irrational. It has continued to ignore some important new scientific evidence about how our genes function. The scientists involved have simply been able to step around the little puddles left by a few Lamarckians. The result is that neo-Darwinian thought has been permitted to adopt a status that is far more entrenched in our world view than mere dogma. It has been elevated to the status of a 'belief' upon which much of twentieth century society has been structured.

Only a few in the last century were brave enough to take on supporters of the neo-Darwinian paradigm, and all have paid the ultimate penalty of being ostracised. However, as genomic and epigenomic data arising from research spanning several scientific fields is analysed, the inadequacy of Neo-Darwinian thought is becoming more evident. As we are entering a century of new biotechnologies, massive amounts of genetic data are starting to create new streams of thought, and showing the way for a major scientific revolution. It is one that will compel us to fundamentally revise our views on evolution and it is inviting us to accept an inconvenient truth: Lamarck was right after all. As with all major scientific revolutions of the past, we are also being forced to alter our views about who we are.

RESURRECTING LAMARCK

For two centuries Lamarck has been ignored, maligned, and ridiculed by popular science writers. Using pictures of monkeys, and the neck of the giraffe, cartoonists have continued to capitalise on his demise. So, like the Copernican revolution, the Lamarckian revolution will represent the end of a long and embittered scientific battle. For two centuries Lamarckian views have been fought tenaciously. Each time the spectre of Lamarckian inheritance has been raised, it has been done so in a completely different scientific and political context. Even to now, two centuries of discrediting has meant that popular science writers, religious instructors, and research funding agencies have all shunned the notion as preposterous and not to be entertained by serious scientists.

There are probably no neo-Darwinians with a higher public profile than Richard Dawkins, a biologist at Oxford University. His views are summarised in the following pronouncement, "It is an established fact that all life on this planet is shaped by Darwinian natural selection". [110] Such is the conviction of some of our most prominent intellectuals in evolutionary thought.

For a century, the neo-Darwinians have been resolute in diverting our focus to an evolutionary theory based on the forces of natural selection and random mutations. It seems that neo-Darwinian thought has been developed and nurtured to now to somehow defend society from the potential 'evil' associated with Lamarckian influences. Some of the more extreme neo-Darwinian commentators have also attempted to link Lamarckian inheritance with Creationism to build a climate of 'guilt by association'. Given the emotive social, political and religious issues that these influences raise, a world run by pure chance is easier to defend than one that might unlock a wider public debate on genetic and epigenetic responsibility.

It seems that the oddest coupling between the forces of natural selection and environmentally directed change needs to be accommodated within the emerging evolutionary paradigm. This is not arguing against either a Lamarckian or a neo-Darwinian view of the world. It is suggesting that the two need to be accommodated within a complex set of algorithms operating on epigenetic and genetic elements. For these arguments to make progress, we need to resurrect Lamarck after two long centuries.

THE EMERGING EVOLUTIONARY PARADIGM

In describing science as a social construct, I mean that it is moulded, framed and placed in a context that is convenient to the most influential scientists of the day, rather than the contributing scientists themselves: it is one of the more splendid artefacts of our imagination. A watered down version that is the most convenient and palatable for the day is then displayed for public consumption. It is moulded to support our beliefs, social values—as well as powerful industrial and political alliances.

In the case of the development of our understanding of the process of evolution, it has frequently been the case that the scientific theories put forward have tended to take on a much wider social significance. This is because our ideas on evolution impact some of our most deeply held beliefs about who we are. The implications at each stage of the growth of our scientific understanding have therefore been extended well beyond the literal scientific context put forward by the original proponents. The result is that the ideological interpolations and political imperatives of the day have outlived the original hypotheses. In the case of neo-Darwinian thought, the idea of natural selection and how it is presented to children in schools bears little resemblance to what Darwin actually wrote and his total contribution to the debate on evolution.

For more than a century we have been told that the basis of what we look like and for our health is pre-determined by our genes which are unaltered during our lifetime. We have built our political, social welfare, economic, education and health systems based on genetic determinism. But now a radically new understanding of how our genes work is beginning to unfold before us. The emerging paradigm is based on natural selection working with heritable feedback loops that act together to update our genome as we rub up against our environment. Both are crucial. A key feature of the emerging paradigm is that every genetic organism exists in a state of what can be best called 'genetic flux', with genetic and epigenetic states dynamically changing in response to environmental feedback. Another important feature is the notion that every enduring system needs constructional constraints: this is an essential requirement for any enduring biological system. This is the essence of the emerging new paradigm. Contrary to popular belief, full-blooded Lamarckism is consistent with neo-Darwinian processes. It is a matter of definition, and understanding the active genetic and molecular mechanisms at play.

Gradually too we are discovering the prescience of the Frenchman Jean-Baptiste Lamarck. If we view evolution as being powered by genomic and epigenomic systems interacting with the environment every sensory moment of our lives, then the origin of complexity disappears.

The timeframes observed, the acquired inheritance effects experimentally verified in our physiology and our cognition, and the possibility of rapid evolution in times of stress all start to become comprehendible.

* * *

Of course the good thing about adopting this view is that it is consistent with the emerging data. Almost all of the associated predictions it makes are, or soon will be, scientifically testable. While we still have much to learn about the highly regulated nature of the informational pathways linking our environment to our genes, we have the technical and conceptual tools to take a much closer look.

* * *

9. EVOLUTIONARY LEARNING LOGIC

All things come out of one, and the one out of all things.

Heraclitus of Ephesus (550-475 BC)

In the previous chapters we looked at some of the molecular mechanisms involved in evolution that have distinctly Lamarckian implications. In the process, we have also discovered the fluid nature of genes. Some include variable regions ready to insert new sequences, and there are other regions that include highly conserved blocks of DNA code that are common to all life. In this section, we look at the logic involved in updating some of the known variable regions of our genes and epigenetic systems by viewing an organism as an information system.

We have seen that our genetic and epigenetic code can be edited using functions similar to those we are familiar with in a word processor. Sequences can be cut, pasted, copied, joined, re-arranged, deleted, edited, and written in a 'write protect' mode. We have also seen that genetic and epigenetic information can be exchanged among different individuals using the range of genetic transfer modes summarized in the previous chapters. The genetic and epigenetic code systems involved are highly distributed throughout our organs and body, and they operate as a well coordinated highly parallel system. Whole sets of instructions at the level of our DNA, RNA, epigenome, cells, tissue, organ or whole body can be simultaneously executed. We also now understand that there are many different types of environmental stimuli that can result in heritable changes at the level of our genome or epigenome.

One of the most intriguing features that we have discovered about a heritable information system is that its elements are essentially re-configurable self-learning codes that learn how to respond to a new environmental signal by relying on just a few steps. If we view these steps as an algorithm, then it can be argued that humans have evolved to rely on the same evolutionary learning logic at all levels. That is, the same evolutionary learning logic can be applied to our genes, to the development of memory, our use of language, and how we build and improve our social structures.

In general terms, a Lamarckian evolutionary learning algorithm operating in an organic information system involves just five basic steps:

First, the system must be able to rely on its sensors to scan the environment, and then signal that a problem might exist.

Second, the system must be able to compare its pre-existing memory patterns with the environmental input signals received by its sensory inputs, and then identify the memory patterns that provide the closest match to it ('best fit').

Third, using the selected 'best fit' with an existing memory pattern as a starting point, the system must be able to initiate a set of actions to alter it so as to create a new and more perfect memory match with what has been sensed as new information.

Fourth, the system must be stable and able to use the improved 'best fit' memory match created to trigger an appropriate and coordinated set of biological responses.

Fifth, the system must be able to update its memory store ready for immediate recognition and response if an identical or similar external information pattern is detected in the future.

There are some critical differences between this Lamarckian evolutionary learning algorithm, and one based on a neo-Darwinian model. It relies on the nature of the signals received as input from the environment to trigger a response. It relies on pre-existing memory systems, and it involves feedback in a form for updating the memory so that the next generation can benefit from any changes.

From a computational perspective, each step in this learning process might require a form of tree branching. When multiple systems are involved, the result is a rise in complexity. This logic has been adopted to gradually increase the complexity of an extraordinary large number of organic systems. It can be used to describe the learning that occurs in our immune system genes. There is some early evidence to suggest that the same algorithm is used by our central nervous system genes. We can also use the same algorithm to describe many of the learning processes we use in our lives. The algorithm can be used to describe how we learn to play chess, how to play the piano, or how to enable our political and economic systems to evolve in harmony as the environment changes.

There are some fascinating examples of important evolutionary processes that have incorporated this approach for adaptive learning in response to environmental stimuli. But let's revisit the immune system genes first, to consider how they use evolutionary learning logic.

EVOLUTIONARY LEARNING LOGIC IN THE IMMUNE SYSTEM

When the immune system comes into contact with a foreign pathogen, identified as 'non-self', it triggers an immune response to ensure that the foreign pathogen does not harm its host. First it calls on its store of antibody genes that are made into a three-dimensional structure using DNA antibody genes. It then identifies those molecules that are the closest three-dimensional match with the three-dimensional surface structures of the foreign pathogen. If a close fit is found that can lock onto the foreign invader and destroy it, then this is multiplied— 'cloned'. The new clones are released to mount an effective response against the now rapidly multiplying invaders.

At the same time, hypermutation processes are triggered to rapidly create a diverse population of modified antibody structures. Each new variant of the pre-existing 'best fit' antibody is tested to see if it provides a 'better fit' to the three-dimensional surface of the foreign antigen. Using a 'survival of the fittest' approach, only the most successful new antibodies are selected and cloned (i.e. those that provide the best three-dimensional fit). This process is repeated until the foreign pathogen is completely eliminated. At each stage, only the best and most effective antibody is selected and rapidly cloned to fight the new invader. It is only these that are used to create new variants.

The immune system uses this learning process over and over again to gradually build a more effective immune response to new infections. It is a strategy that also enables the immune system to learn to fight infections that genetically mutate while attacking the host, in an effort to outsmart the immune system. In this case, the reiterative process of finding a new 'best fit' is used to keep up with, and eventually outpace the mutating invader.

When the immune system has effectively eliminated the foreign pathogen, it retains a very low titre of the most successful new antibodies created. The updated information is reverse transcribed into our DNA for future use. The updated antibody genes then serve as a more permanent back-up copy of the new successful antibodies created during an immune response.

LEARNING TO SMELL THE CHEESE

Perhaps one of the best examples of how evolutionary learning logic is used to update our genes is found in the odorant receptor genes—the genes that allow us to differentiate Chanel No. 5 from flatulence even if we're blindfolded. Recognition and response of this large family of genes involves a very different set of processes to those used for recognition and response of a foreign antigen in the immune system. However, the evidence we have suggests that the evolutionary development of this family of genes has relied on the same evolutionary learning logic.

One of the best descriptions of how we use our sensory inputs and information flow in a neural network is provided by the Nobel Prize winners Linda Buck and Richard Axel. Buck and Axel were awarded the Nobel Prize in Physiology or Medicine in 2004 for their work on odorant receptors and the organization of the olfactory recognition and response system. Their research has led us to an understanding of why we can recognize and remember about ten thousand different smells with only three hundred and fifty different types of odorant receptors.

Buck and Axel discovered that a large gene family of around one thousand different genes is involved, representing around three percent of our genome. Each gene gives rise to a different type of olfactory receptor. Each receptor is highly specialized and can detect only a few different odours. While smell is absolutely important to mammalian pups or mice, it has become less important for humans. Mice use around one thousand different receptors, while humans use only three hundred and fifty. This bears testimony to the importance of smell recognition in our evolutionary past.

The olfactory receptors are all located in a small area in the lining of our nasal cavity, and they detect inhaled odorant molecules. Once an odour is detected, a thin nerve signal is sent to a carefully mapped micro domain ('glomerus') in the olfactory bulb in the brain. Each type of olfactory receptor sends signals to a specific micro domain. That is, there is a one-to-one mapping between each type of receptor and a single micro domain in the brain. All of the signals reaching the olfactory bulb are then processed and sent to other parts of the brain where single neurones are able to accept more than one type of signal input. Here the information from several olfactory receptors is combined to form a particular pattern to recognise a new odour. Once processed by association with previous memories, thousands of possible odorant patterns become recognisable by us. The signals are also relayed to other areas of the brain so that we can link images, sounds or other sensory inputs associated with the new olfactory signal. When we smell mint, we associate the odour with a mint leaf. We are then able to consciously develop our experience of different smells of cheeses, wines or flowers through this complex system of pattern recognition and learning by creating and remembering new patterns.

The odorant receptors use this process so that we are able to link an instinctual, emotional or physical response to the sensory input. While we do not understand all of the steps involved, the new memory patterns are stored for future reference, and some are inherited. An example of some inherited odorant recognition patterns is that associated with pheromones that are linked to instinctive reproductive behaviour. In the case of dogs, even a small amount of sex pheromones will sexually arouse a nearby male dog. Dogs have evolved to use several odours for basic communication when greeting each other. One sniff enables them to tell the gender, age, emotional state and their reproductive stage. Using urine marks, they can identify an individual dog. They also overmark the scent of other dogs of the same sex. A female standing near another urinating female will often wait until they leave, and then go over and immediately urinate on the same spot! All of these instinctual actions rely on the ability to learn and remember a whole range of different scents.

GENES LEARNING TO TAKE ON A NEW ROLE

The MHC ('major histocompatibility complex') genes provide an example of how a family of genes can evolve from providing a specific function at the molecular level, to also being able to provide a similar support role for a whole organism.

In the immune system, the MHC genes play a key role in enabling T cells to distinguish between self and cells containing foreign viruses. To do this they rely on three-dimensional surface structures of a virus infected cell. They compare these to a pre-existing series of MHC gene markers to help the T cells discriminate between self and foreign viruses. Their role in evolutionary terms is dependent on a capacity to update existing memory as environmental signals change.

At some stage in our evolutionary past, this family of genes adapted to assist the whole organism to use its olfactory system to distinguish between different MHC types. Many of the odorant receptor genes are in the MHC (histocompatability complex) region of our genome. The genetic markers that enable T cells to recognise self from non-self are also in the MHC region. This enables female mice to use pheromone odorants to distinguish between male relatives who have similar MHC molecules, and their preferred mating partners with different MHC molecules. This ability to distinguish between self and non-self at the level of the whole organism is a learned phenomenon.

Some of the MHC molecules are excreted in the urine. The result is that females can benefit from the well known phenomenon of hybrid vigour by mating with unrelated partners. The MHC genes also enable the cats and dogs of the world to mark their territory so that those who sniff it will be able to distinguish between relatives, pride or pack members and foreign animals.

Thus, this very large family of MHC genes includes markers that have been adapted for two different purposes. One is in the immune system that operates at a molecular level, and the other for the olfactory system that operates at the level of a whole organism.

By discovering the complex relationships between genes within a gene family, their function and their relationships with other parts of our bodies we are starting to understand just how we have evolved as a fully coordinated and responsive organism by re-using the same evolutionary learning algorithm. In the process of understanding this we are also drawing previously unrelated fields in biology closer together. The idea that one gene family performs only one function is no longer valid. Similarly, the idea of a single domain for perception and making decisions no longer holds true.

A LAMARCKIAN VIEW OF LEARNING LANGUAGE

If we have evolved using evolutionary learning logic, then it is no surprise that the language we use to communicate is also evolving using the same logic. When we break our language down into each word or phrase, we have learnt to associate it with a 'pattern of thought' that gives it meaning. It doesn't matter if the word represents an object or an abstract idea, we will each attach a slightly different meaning to it as meaning relies directly on pre-existing memory patterns. Words then become labels for our thoughts so that we can communicate with each other.

So perfectly developed is our language that we are able to take meaning from these pages as we read, just as a protein makes sense of an RNA sequence. The protein knows which direction to zip along. It knows what to read, and it has the ability to go back, check, and then edit. This is what you do when you read or write. You use letters grouped as words. The words are grouped into sentences. The sentences are then grouped into paragraphs. Enough paragraphs carefully crafted, and you may end up with a book that has a very specific purpose. In any book, you re-use the same words over and over, but each sentence is unique.

We use the same rules to communicate with language as the genome uses to make efficient use of the large number of gene sequences it has at its disposal. If we need a new word to link our thoughts to a new object or concept, we make one up. It may or may not become adopted for wider use. If the genome needs new genetic information, it creates it. But when this does occur, there are a lot of regulatory systems present to ensure that our genes do not become overrun with too much new information, and to ensure that only updates that are required enter our genes. Our social groupings also have enough inbuilt regulatory processes to prevent useless words from permanently entering our vocabulary, or too many at once. This means that our language is evolving using the same evolutionary learning algorithm as that used by some families of genes.

Our self-learning processes have also evolved to rely on the same evolutionary learning principles. Each time we plan for the future, or respond to a compliment from a friend, we consciously and quickly play a range of possible responses. We then select a 'best fit' outcome based on the options available. The 'best fit' scenario used in response is usually one that aligns our beliefs, our own self-identity, and suits the social context.

Based on these principles, learning is a gradual process that is altered by experiences that are stored as molecular memories. It enables us to interpret the events in the environment around us. When we explore the environment, we are constrained by what we have learnt previously. That is, we are only able to explore things in our environment that present meaningful possibilities.

Right from the time we are born we start to build up our experiential memory. Just as a small single celled life form responds to vibrations, changes in light intensity or changes in temperature, so too did we when we were first born. Later, we learn to crawl, to interact with those around us, and then walk. Each stage of our learning required constant environmental feedback about our progress.

When a small child first sees a dog in a street, they see it as simply a dog that barks. As the child grows older, the dog becomes a male Jack Russell called Astro. Over time, the child's knowledge of the same Jack Russell becomes associated with additional information and emotions. Images of Astro as a favourite family pet, and as a hardy, fast and fearsome little dog are built up. All of the additional information is linked to the original image of a small dog that is registered on the visual cortex of the child's brain. At each stage of learning, the child relies on environmental feedback to form an updated memory pattern.

Sometimes when we interact with our environment, associations with other memories may trigger an involuntary response. Responses to environmental triggers also grow stronger through use. When we first fall madly in love with what we believe is our ideal mate,

the resulting peptide flow dominates our eating, our
sleep patterns and our dream states. Every moment we
are awake we are waiting to receive triggers associated
with that person. Smells, music, or images in a crowded
subway may trigger associations. When we meet what we
think is our perfect mate we may also engage in some
rather contradictory self-talk, "Don't worry about what
he is thinking about me, he is too busy worrying about
what I am thinking about him." It is evident to any of us
that have ever been in this state that we enter into a new
phase of self-learning using the evolutionary learning
algorithm.

When you go to sleep, this same evolutionary learning
algorithm continues. You might set about planning the
next day in a dream that plays out some unexpected
scenarios. Over and over again, you throw up a range of
possibilities arising from your experiences, and you use
dream scenarios to test their 'fit' until you end up with
one that you are completely happy with. This one you
remember and implement.

The same processes are involved in acquiring
complex ideas linked to a large amount of associated
information. For some individuals, a scientific problem
may be solved with an 'aha' moment while driving a car,
or sleeping, as they subconsciously run through some
possible scenarios. But once we recognise a 'best fit'
solution, we use it, and remember it for the future. We
may also re-use it to improve our ability to solve other
problems. Just as we remember patterns of behaviour,
we also remember complex ideas as patterns with a
series of complex links to other information.

We all act like this because that is how we are
designed to function from our genome up. In the future

we may uncover all of the molecular pathways involved. All we know now is that these learning behaviours bear some uncanny similarities with the algorithms that were developed millions of years ago to solve problems by learning at a molecular level.

In every learning system, erroneous or false information does sometimes manage to filter through the complex regulatory systems. The same is true for our genomic and epigenomic systems, our protein folding patterns, our central nervous system, our own memory, and the memory of the scientific community. Extra rules for restricting both the quantity and type of new information that becomes integrated into a permanent memory are beneficial to ensure the stability of a system. Such is the nature and implications for all living systems.

Yet sometimes too this can be a weakness. By being the animal at the top of the evolutionary tree, reliant on such complex organisational and learning structures, we could be slow to respond to possible deleterious threats. One example is the debate about climate change. Some scientists and governments cannot agree that it is real, or on its impacts, its cause or the time-frames involved. Yet, just suppose that the threat is real and that we only have five years to respond before a rapid and catastrophic climate change is triggered by reaching a threshold set of conditions not previously understood! Given the layers of complexity making up our nature, we are simply not able to agree on whether or not a threat has been properly identified, let alone agree on how to respond quickly. In the event of this scenario, millions of lives could be lost. We are lucky that such an event is not likely to make us extinct as a species.

LEARNING AND INSTINCT

When learning characteristics are used frequently by a species, they become labelled as 'instincts', and these are believed to be hardwired into our genome. The idea that our 'instincts' provide evidence for acquired inheritance is also often proposed to support a Lamarckian view of evolution.

A theory put forward by James Mark Baldwin in 1896 invoked the idea that the ability of individuals to learn in their lifetime played a role in guiding the evolutionary process. Thus, the idea that learning characteristics might be inheritable is now sometimes referred to as the 'Baldwin effect'. In extending this idea, Baldwin postulated that evolution eventually replaces abilities that initially required learning. He believed that learnt behaviours may develop into what he called 'instinctual behaviours'. Baldwin viewed his ideas as a way of enabling contemporary Lamarckian and Darwinian views to be accommodated in a mutually supportive way.

In any population, the individuals who learn the fastest will have obvious survival advantages for gathering food, evading predators, and building up social structures. As time goes on the capacity within a population to learn will increase. But at some point, some repeated behaviours become instinct, and the individuals of a population are then able to focus on learning a new set of behaviours. While, the idea of the Baldwin effect is not given much credence today, our understanding of evolutionary learning logic is making it possible for us to develop new explanations for the Lamarckian-like effects he described.

LEARNING AND INTELLIGENCE

If we think of ourselves as having a biological basis for the way that we go about solving problems, then it also provides us with some new ways to think about our own intelligence. So, in the context of the new genetics, what is 'intelligence'?

As the debate in our schools continues on how to improve and measure 'intelligence', it seems that the very idea of what is an intelligent child is constantly changing. The social and political winds are continuing to shift the sands. Yet, it is generally agreed that any definition of intelligence cannot ignore the environment, and that any measure of intelligence must provide a relative quantitative measure of how well an individual solves problems when interacting with the environment.

Our society has many ways to measure intelligence. Measures of academic performance are based on intelligence activities and intelligently organized responses over time. A capitalist system allows for rewards to be taken by individuals who have good financial intelligence. We reward scientists with prizes for providing us with new knowledge. We reward artists, athletes and other areas of human endeavour with money, prizes, recognition and other privileges. In almost every case, a degree of innovation is required to create some new form of competitive advantage.

Intelligence then, can be considered to be a phenomenon based on physical factors. Inherent in this view is the assumption that the mind cannot be considered as separate from the body and the environment. It is not abstract. It is the sum total of our ability to use our senses to scan our environment and make some sense of it in terms of what we already know. We each evaluate our environment in terms of what knowledge we have accumulated over time. In that context we can quickly determine what is abnormal. We ask, "What is different, but is a good match with what we know?"

This evolutionary learning process allows us to take our time, to scan the environment, integrate and evaluate a number of related observations. We can then mount a pre-meditated and organized response if we choose. This also means that we are all capable of increasing our intelligence. While this might seem like a triumph in reductionist theory, it should be remembered that the actual processes at play at every level are incredibly co-dependent. Many steps, and proteins, RNAs, the epigenome and the genome are involved. While we cannot yet describe all of these, we are beginning to see some patterns emerging that can be used to make some testable predictions. Our genes then set our own internal agenda, while our thinking and learning build on this complexity in ways we have yet to fully understand.

LAMARCKISM AND SOCIETAL CHANGE

Over the years, many have argued that our social evolution is Lamarckian in nature. In 1959, Peter Medawar recognised that cultural evolution is a phenomenon that is similar to Lamarckian evolution. Other proponents included the philosopher Karl Popper who became one of the most admired philosophers of the Twentieth Century. A Lamarckian theme is omnipresent in many of his works published in the 1960s, and early 1970s. Decades later we can support

Popper's views by noting that our social structures evolve using the same evolutionary learning logic as that used by some molecular systems to adapt to change and to grow in complexity in a coordinated manner.

As a society we have used evolutionary learning logic to improve our individual intelligence and collective knowledge base. Educational institutions like schools and universities have been established using the same learning logic. We are constantly scanning our environment to look for methods to improve the effectiveness of these learning systems, and to ensure that they are structured and updated to meet new environmental challenges.

We have also developed increasingly sophisticated regulatory and feedback systems to enhance the effectiveness of our ability to make group decisions. When a government is given a problem to solve for society, they use bureaucratic resources and processes to identify possible solutions. These are evaluated using a range of methods, including experts in the field, debate and public feedback. The debates are an important part of the evaluation process. Sometimes they are outrageous and at times the options being considered are just simply ill-informed. Yet we trust and respect the parliamentary process to make laws for our society. Finally, a 'best fit' solution is identified and adopted. If it is an issue that needs to be managed on an ongoing basis, then the new 'best fit' solution becomes enshrined in legislation. The more permanent legislation can be compared to some critical sets of RNA code becoming reverse transcribed into DNA that are securely stored in the nucleus. The legislation and the DNA are both forms of a more permanent memory store of new information.

This is a process that many accept for making important group decisions because it is a natural extension of how we function as an individual and as a species. In fact, our own ability to use evolutionary learning logic is so much a part of our form and its functions, that it is quite difficult to imagine adopting alternate ways of learning or making group decisions. This is an interesting point, because it challenges us to conceive of new ways to evolve a more effective group learning process based on our existing biology.

Science also adopts the same processes for considering what new research projects are to be funded. Those that are a 'good fit' with what is known become more readily accepted. Sometimes, wide acceptance for an idea by peers is used as a justification for providing additional research grants to related research proposals. However, once an area becomes fully characterised, and it is aligned with social, economic and political goals, then it may become more widely accepted as 'a truth'. The enshrining of new scientific knowledge is done by awarding prizes, and by perpetrating the new knowledge in our text books as 'fact'. Recording the new knowledge into more permanent memories such as school text books is the final step in creating a new scientific 'truth'.

In all of the previous examples, the key to unlocking our ability to learn by evolutionary learning logic is the ability to incorporate feedback, and to form associations among learning systems. As we evolve, the process of augmentation through increasing complexity relies on association and feedback at all levels. At every level, we use the same fundamental processes to search for new information, match it (if possible), and update the memory store to accommodate new information. It has been happening in our genes for millions of years.

It is therefore axiomatic that in a biological system, no multi-step process is carried out by chance. We have gradually built a repertoire of associated forms of memory over evolutionary timeframes. These form the foundations for future evolutionary learning.

Over long time-frames, endless repetition and repeat reinforcement of certain behaviours through generations will result in some common instincts being passed through many generations. Karl Jung, the Swiss psychiatrist referred to this phenomenon as a 'collective conscience' that is present in us all. In a landmark paper entitled *A New Factor in Evolution* (1896), he described the collective conscience of mankind as consisting of deeply engraved behaviours associated with experiences that have made their way into the human psychic makeup. These later became known as 'Jungian archetypes'.

Jung described collective consciousness as a level of consciousness common to us all from birth, and not dependent on learnt behaviour or experience. Although we have some individual variations, we all harbour a capacity for love for each other, love for our children and sexual attraction. We are also born with a repertoire of images that make most of us go into an immediate fight or flight reaction such as when we see a snake, or a spider. Experiences like this allow us to reinforce and augment our instinctual repertoire.

The example of our reaction to looking down from a great height is an interesting one. How did nearly all of us come to have the feeling of 'falling' when we look down from the top of a tall building, a ladder or a high bridge? While we can learn to overcome this fear of falling, the 'fear' is actually a feeling of the shock of falling freely through the air. At least that is what our subconscious associates it with. When our visual cortex interprets an image from a great height to the ground below, it associates the feeling of falling with it.

Perhaps this instinctual behaviour arose from a time when our survival depended on swinging safely through the trees. For the association of the feeling of falling to exist with heights, we must have had ancestors who lived to remember this feeling and pass it on to offspring. After all, we don't have the same feelings triggered when we look across a desert plain, or at a plane flying in the sky. It only occurs when we look down a distance that is much greater than our own height. Jung claimed that endless repetition and repeat enforcement through generations have deeply engraved the associated experiences into our psychic makeup. He believed that this type of instinct is acquired through Lamarckian mechanisms, so that learning could eventually become inherited.

Through association we link everything in our environment with our memories, our individual knowledge base whether it is conscious or unconscious, and our collective knowledge systems. It also implies that the thought processes of learning, language, creativity, instinct and our own consciousness become processes relying almost instantaneously on the molecular elements linking our form and our genome. It means that we give our thoughts meaning by relying on what is already programmed in the underlying networks of associated memories. As we learn we are capable of updating them.

This is one amazing code for life and living! Our own evolutionary processes are designed to exploit this biological code in a most efficient manner. There is no artificial intelligence program that comes close to producing such thought processes, nor the capacity for learning that we harness. This process has also acted to restrict the speed at which we can learn. At every level the admission of new information onto a pre-existing memory store is highly regulated. If too much new information was permitted to enter a system, then the whole system could rapidly become unstable.

The idea that the evolution of our own learning such as the use of language, and our social systems is Lamarckian and somehow linked to our genes is not new. Yet, we have been very slow to admit that in between there might also be Lamarckian feedback loops operating to enhance learning at a molecular level. In the context of the new genetics, our genome, the immune system, our central nervous systems genes and possibly others, our language and society are all structured to use and re-use some basic evolutionary learning principles. The key physical difference in each system is in how information is represented, how memory is retained, how diversity is generated and how the learning is used.

But how has evolutionary logic given rise to human consciousness?

WHAT MAKES US HUMAN?

While I have no intention of speculating on the origin of life itself, I find that we are now at a stage where we are at least beginning to understand the enormous potential locked inside the amazing package of code that we come with. It develops as an ever-evolving interactive system of information that is able to replicate itself. It is able to generate diversity for natural selection to act on, and to improve pre-existing code for making certain physical structures and for regulating the use of the code itself. It includes a living library consisting of a record of many different types of experiences, including antigenic experiences, central nervous system responses, and the memory of our daily interactions. Ultimately it has given rise to our consciousness. It is our consciousness that makes it possible to choose how we respond to almost any unexpected events. It also provides us with enough insight to plan for our future.

But what does it mean to be a fully conscious human being in the context of the new genetics?

To even begin to answer this question, the first thing we are coming to realise is that we must adopt a holistic view of ourselves. Nowhere is this more evident than when we take a look at evolutionary learning logic, and how it has enabled us to develop our own unique 'sense of self'. It is this that truly makes us human.

From a computational perspective, the human body has around 100,000,000,000 neurones. Each neurone has anything from one hundred to one thousand connectors. Some of these are clumped together in the autonomic nervous system responsible for our instinctual responses. Others are linked to our inherited primary emotions. Others are in domains dedicated to deciphering sensory inputs. Some are for coordinating motor outputs, and a large group are for performing higher cognitive functions. Many of the basic physical and biochemical interactions are becoming quite well understood. The co-evolution of all of these physical characteristics has enabled us to become extremely dexterous and cognitive as we interact with our environment. But the most fascinating organ that our evolution has given rise to that truly makes us different from other species is the human brain. So let's take a look at this special organ.

The human brain has over a trillion processors, each working at a few hundred operations per second, making it capable of performing around 10^{13} to 10^{15} operations per second. Powered by around twenty five watts of energy, it is also endowed with rich connectivity between the tiny processors making the number of possible permutations and combinations of network paths almost infinite. This connectivity and processing power makes the human brain suited to solving complex and often abstract problems. It is essentially an analogue device able to solve three-dimensional problems, and at the same time act as a digital information system able to solve mathematical problems. As a massively parallel processor of information being received and instruction sets being sent, it is difficult to recreate its complexity.

The idea that the brain has evolved as a set of co-located independent modules, each with a separate function therefore seems to be quite irrational. Each section has evolved in cooperation with each of the other parts. They act as a carefully networked whole, whose functions are fully integrated. Its normal operation relies on the rest of the body. Regulation, sensory inputs, nutrition and protection are just some of the body functions that need to be coordinated and networked in a fully functioning brain.

Our brain is also able to respond to changed environmental conditions with incredible plasticity. With over one hundred billion neurones in our brain, a newborn baby has huge potential. However, by the time a child is two years old, it is estimated that the number of brain connections has decreased by around forty percent. The links it does not need are simply removed. The result is a more efficient brain that has been adapted to suit its environment. As an adult, when we are placed in a stimulating environment, the brain suddenly increases the number of connections again, before deleting unused ones in later months.

When we look at the human brain as a layered structure, we see our past. The structure of the brain reveals how additional layers of complexity have been built up from the old reptilian complex at the base of our brain that controls body regulation, through to the neocortex module at the top for planning our future. Essentially the human brain consists of three basic modules, each of which has been augmented over time.

The first is the reptilian complex found at the base of our brain. All of our strong instinctual urges are controlled by the remnants of a small reptilian brain from our deep past. Its main goal is survival. It is connected to the spinal cord and all of the nerves extending from it to the various body parts and organs. It regulates all autonomic nervous system processes. The heart keeps pumping as it is prepared for immediate fight or flight responses when we receive a shock, when there is no time to think about what is happening, or when we don't know what the best response might be. Some jump when they see a snake, and adrenaline pumps through the body. We devour food when starved. Although we are conscious of these processes, we are also aware that there are often few logical connections between spontaneous urges and our logical thought processes. Over time, its functions and control have been augmented by adding another brain on top of the primal one.

In the middle, the mammalian, or mid-brain (limbic system) is hard packed and connected strongly to learning and emotion. This middle brain development stage has enabled us to consciously create social structures and a degree of self awareness. It has enabled automatic learning based on association. It enables us to use the environment for survival. Memories start to build up, and these act as a force on how we interpret what our senses pick up. The visual cortex at the back of the brain imprints images through association with previously known colours, patterns and forms. Sometimes these are also associated with other sensory inputs such as smell, feelings, or sound. And sometimes these are associated with deeply imprinted feelings of fear, anxiety or happiness.

This means that we can only see what we believe is possible according to associated patterns in our memory, and all that we see has an emotional aspect to it. In other words we go back in time, every time we see, hear, smell, taste or feel something. For example, great art serves as powerful visual symbols for our emotions. Artists like the Spanish painter and sculptor Pablo Picasso are masters at creating images to communicate with us by drawing on our deepest emotions, moods and feelings.

The third brain is our forebrain. It provides us with the ability to become self aware thinkers capable of forethought and creativity. These capabilities provide us with the ability to develop new insights into our own view of evolution. It provides is with an awareness of our inner self, and the ability to create a new future based on this awareness. The forebrain is therefore essential to our idea of 'self' and what it means to be human. Emotions, beliefs, physical appearance, sex and social status are just some of the things that define our sense of self.

The forebrain also provides us with the ability to develop morals, and insight into conceptual problems. Our impulses are constrained under the burden of morality. In turn, our morals determine most of what we do, and why we do it. Our concept of self also changes as we receive and integrate feedback from our environment. We welcome a baby into the world. We listen to what our children really think of us. We see an X-ray showing three separate tumours on our left lung. In each case, the environmental feedback fundamentally alters our idea of self.

An important forebrain function that differentiates us as humans is the ability to guide our most deeply held beliefs. Choosing to live by one's beliefs is perhaps one of the most significant advances in our evolutionary development. Many of our thoughts are influenced by beliefs. Our beliefs about materialism, family, religion, evolutionary thought and politics all help us to make decisions. We constantly work to manoeuvre ourselves into a position which sits most comfortably with our beliefs. This helps us to develop our own set of realities, and in turn to support and reaffirm the positions we take. This is also how we become conditioned to our environment.

Although humans have developed a high level of self awareness, and all live their lives according to a set of diverse 'beliefs', much of our consciousness is still driven by strong and primitive urges. To fear, to love and protect our children, to eat, or to sleep are all so strong that most of us do not question them. The point I am making is that as humans, most of our behaviours are still primitive, and that although we have developed a high level of consciousness, it takes a strong person to be able to control all of these instinctual urges according to our beliefs. Yet we also have the ability to watch ourselves as the thinker conceiving of a new world.

From a molecular point of view, we know that RNA editing is prominent in the brain. For every interaction with the environment and every new thought we either rely on previous experiences, our learning, or both. In all cases we receive and process a range of individual sensory inputs. All of these processes rely on rapid RNA editing. Sensory input can come from touch, visual images, audible sounds, smells, tastes or physical movements such as vibrations or movements. Our brain processes the information by associating it with previous sensory inputs. By relating each individual sensory input with related sensory inputs, we can build up our image network. For example, you hear a sound. It is a voice. You link that voice signal to the image of a person you know. You immediately associate the voice and image with a set of feelings associated with the relationship that you have with that person. You overlay the voice message and tonal inputs. You then decide to react quickly as this person is angry at you, when they haven't been in the past! As the response is to be guided by your own consciousness, RNA editing activity in the brain is raised to a high level during the encounter. As you respond, you are also actively laying down new memories based on the experience. This also involves rapid RNA editing in the brain. It is done quickly to ensure that the anger, the reason for the anger and the feeling that the person's anger generated are linked and that they are all associated with another memory of your relationship with that person.

While this interaction may only take a few seconds, the impact on your memory may become permanent. Learning associated with emotions alters our gene expression, and it gives rise to different forms of thinking and behaviour. The more emotional the encounter is, the greater the likelihood that you will remember it forever.

Our emotions also play another role. Emotions that are common to all, work to bring us together. Our emotions are based on a pattern of peptides released that match a feeling. Strong memories are associated with each pattern of peptide release. It is therefore easy to understand how someone is feeling when they are in love, jealous, happy, or profoundly sad at the loss of someone close. But only once we have had a similar experience. These events draw on the same pattern of peptides giving rise to our emotional capacity. Communicating with another person on an emotional level therefore draws us closer together.

In comparison, the ideas and perceptions we have developed by building on our emotions become highly diversified across a population. They act to draw us apart. In the relaxed ambience of a dinner party, we discover just how differently other people think about politics, new cars, the global financial crisis or global warming. These are interesting topics to discuss among friends. To engage in these discussions requires us to express our perceptions based on the information that is available at the time, and how that rests with our own inner beliefs. It is also the point at which we become highly differentiated through perceptions that draw us apart. Sometimes these discussions rapidly escalate into a heated debate! The message is to rely on emotions if you want to stay close.

Our idea of what makes us human must therefore be fundamentally altered to incorporate a more general and holistic view of ourselves. Our thoughts are basically built on perceptual processes. Our perceptual inputs from the environment provide the main feedback. Our ability to manage environmental feedback and our perceptual responses together form the molecular basis for cognition.

If we accept this logic, then our idea of self, our consciousness and the structure of reason itself all have a biological basis that has been developed over millions of years using Lamarckian evolutionary learning logic at each stage.

* * *

As I stand and gaze into Alice's Looking Glass again, I am confronted with another image: I see a future in which our destiny is limited to becoming faster and more effective at using the same evolutionary learning logic that created us! I gaze a moment longer. How can we escape?

* * *

10. CONSCIOUS EVOLUTION

*"Humanity is going to require a new way of thinking
if it is to survive."*

Albert Einstein

In the preceding chapters, the conceptual objects
arising from the new genetics and their accustomed
place in the neo-Darwinian world have been re-arranged
into a Lamarckian framework. While the focus has been
on understanding what the new genetics means for our
understanding of evolutionary processes, the emerging
more complex patterns have given rise to the concept
of consciously directed evolution. It is this that evokes
some new explanations for the human mind. I therefore
devote this last chapter to introducing some of the
many consequences that flow quite naturally from the
emerging evolutionary paradigm.

Although brief, the issues touched on affect some of
our most deeply held beliefs—and they draw each of us
closer to understanding our own existence.

EVOLUTIONARY THOUGHT AND CONSCIOUS EVOLUTION

The idea of conscious evolution is not new. Scientists
and philosophers such as the English evolutionary
biologist Julian Huxley, the French biologist and
philosopher Teilhard de Chardin and others have helped
to form our early understanding of conscious evolution.
Originally it was considered as a spiritual phenomenon
or an overarching meta-discipline drawing on a number
of normally disparate fields of inquiry. What is relatively
new is the body of scientific evidence suggesting that

it is possible. By developing our understanding of the molecular mechanisms underlying Lamarckian evolutionary processes, we are introducing the idea of conscious evolution into mainstream science.

As we confront the growing amount of molecular and genetic data showing that we are designed to adapt to our environment, sooner or later we will need to confront the personal and global responsibilities of consciously directed evolution: the ability to consciously direct our own evolutionary destiny. Through the conscious choices we make, we can alter the pace and direction of our evolution. This realisation has the potential to propel our mass consciousness to another level altogether. In this chapter we look at just a few of the many implications arising from this realisation.

Our understanding of evolutionary processes can have a profound impact on each of us. It is at the core of who we think we are. In a society adopting a survival of the fittest mantra, there is little or no genetic responsibility of individuals for the next generation. Our political and welfare systems, our international aid programs, our educational systems are all based on this premise. So when researchers verified that DNA was the heritable material that they had been searching for, it was only natural to redefine ourselves in a new form derived from DNA. We also updated our description of how evolution occurred within the prevailing neo-Darwinian paradigm.

Later in the 1980s when we first became aware that our DNA may be updated as we respond to infectious diseases, we modified these views only slightly to accommodate the new findings about the relationships between our health and our genes. But by the mid-1990s these views started to unravel. New genetic analyses were showing that our DNA can be altered by a wide range of environmental challenges such as emotional trauma, lifestyle choices like smoking and depression. By this stage, we started to become aware that our own DNA mirrored at least some of the changes occurring in our life. We were discovering that our actions and our genes were more closely linked than we ever thought possible in the past.

Over the last few years, we have continued to throw up much more evidence to support an even bigger shift in evolutionary thought. Not only are we discovering more evidence to suggest that our pangenome mirrors our life changes from the moment we are conceived until death, but that many of these changes are heritable. In the process of developing our understanding of how epigenetic systems work, we have also had to update our definition of what constitutes heritable information to include both genomic and epigenomic elements, RNA and some protein structures. Like pieces in a giant jigsaw puzzle, genetic and epigenetic data are gradually enabling us to see a new image of ourselves as a highly mutable living organism with the power to direct our future evolution. This new image is delivered in the form of an unexpected realisation: Mother Nature is no longer at the wheel. We are.

Another realisation is that the more pieces of the genomic jigsaw that scientists place in front of us, the more important *our* role becomes.

While science that looks way beyond ourselves teaches us that we might just be 'cosmic accidents' after all, the new genetics is teaching us to look more carefully at our own significance, and our own powers within the grand scheme of things. To live with the knowledge of how our body works to create a unique molecular template for life means that our main form of identity becomes focused on an internal state. That is, our relationship with the environment will depend more on our awareness of changes occurring within, rather than on our external physical attributes.

The precepts of conscious evolution therefore lie in our understanding of how the environment and our lifestyle choices impact our inheritable material. It requires a vision of how we want to be, and what type of offspring we prefer.

What is arising is an opportunity to drive further evolutionary augmentation and greater diversity of us as a species. Never before have we had such a choice. Yet, to define a new vision so that we can complete our own image in the giant jigsaw puzzle of life is a new journey that we are only just beginning.

Philosophers love this sort of dilemma. Living through a paradigm shift of this magnitude, and spanning so many disciplines could potentially have a huge impact on how we perceive ourselves, our society and our moral codes. The very idea of the space called philosophy is expanded. We have far more conceptual realities and possibilities to fit into our philosophical prisms. While further research will develop our understanding of the underlying molecular processes, it will not help us to define who we want to become. Put simply, it is a paradigm shift that will leave us all wondering about the potential of our genome.

GENETIC RESPONSIBILITY

Another realisation is that the idea of conscious evolution awakens a sense of genetic responsibility.

As we become aware of these responsibilities, the real challenge will be to find a balance between our lifestyle, our individual rights and our responsibility to others and future generations. There will be many more opinions than those that exist today when discussing controversial issues like how to regulate stem cell research.

While we are still a long way from understanding the details of the molecular and genetic processes involved in adaptive evolution, we are already using many technologies that will influence our evolutionary direction. We are influencing our evolutionary future by creating and selecting test tube babies, and by developing new drugs to control moods. We can choose to block out horrible memories for emergency medics caring for car accident victims using a drug called propranolol to reduce post-traumatic stress. While it is normally prescribed to control hypertension, it is also known to interfere with the natural processes involved in re-creating memories. We can enhance our own memory. We can condition buying habits and increase consumption by adding hypersonic messages to television advertisements or drink vending machines. All of these can give rise to new molecular records of thought patterns. The new genetics is delivering to us a molecular understanding of how these technologies can be used to direct our evolutionary future. It is this new awareness that is giving rise to our ability to influence our evolutionary future.

Every living moment our body interacts with the epigenome and the genome. Each reflection of this interaction has a molecular representation. At some point you might pass some of these representations to the next generation. Just like our other organs and our immune systems have evolved to become more diversified and unified over time, our idea of consciousness has also become more unified. We are rapidly approaching a new evolutionary juncture where we can analyse our own consciousness in scientific terms.

Understanding consciousness as a physical property of all complex living things means that consciousness will no longer be a 'scientific mystery': the form that our consciousness takes on has emerged quite naturally from both Lamarckian and neo-Darwinian evolutionary forces. It is an implicate part of our physiology. I find this view consistent with the emerging scientific facts. The Soma is awakened!

Yet, the range of possible future outcomes remains incomprehensible, even pointless. So what does the idea of genetic responsibility mean now that we can go one step further and potentially create another layer of augmentation to our physical and conscious existence? Where will it lead us?

INTRODUCING HOMOSAP2

Right now, as we contemplate what our evolutionary future might be, we are all actively writing the code for HomoSap2. Simply by being aware of the processes involved and the possibilities that this might lead to, we are laying down an early blueprint for HomoSap2. Even if we do not have children, we are influencing others in ways that we have not previously contemplated.

One of the most immediate things that HomoSap2 will have to learn is how to domesticate the genetic genie now being released in research laboratories around the world. With gene transfer technologies, genetic feedback loops, genes that move around the genome, rapid whole genome screening and new genetic engineering technologies to work with, HomoSap2 has a lot of work ahead of him in the next fifty years.

In the new biotech century, HomoSap2 will realise that to understand his new world, he will need to think of genes as just a part of a whole living system. There is now no doubt that our fate is determined by our pangenome, the environment, stochastic development events and the will we exert to direct our behaviour. A reductionist view that focuses on just the genome, the epigenome, a single metabolic pathway or an organ will not be of much help to HomoSap2.

HomoSap2 will need to think of genetics more in terms of the algorithms that nature has created to activate complex learning pathways. He will need to view his existence and his own consciousness as just a part of a rather complicated and dynamic living system whose somatic form follows function through interacting with the environment.

But who is HomoSap2?

To answer this question, there are an almost infinite number of possibilities. Some are based on optimism, others on a pessimistic view.

According to the optimists there is much promise in being able to consciously direct our evolution to lead to a new form of utopia. Of course, we can only ever strive for utopia. How we manage the responsibility of conscious evolution in a changing society will be shaped by our collective ideas on utopia. It will determine our destiny.

The word 'utopia' was first used by Thomas Moore to explain his own ideas of what a perfect society might be like. In his book *Utopia* (1516) he describes his ideal view of towns, marriage, life, religion, the military and the other things that were important to his society. In a world where the actions of scientists and our own intelligence act to increase our self-awareness, there are many more possibilities for creating a new utopia. Some live their whole lives seeking fortune, drugs, fame, physical beauty and almost endless other pursuits. But as we become the custodians of our own evolution, these possibilities might only be temporary.

So how do we address the pessimists?

Further scientific research will need to be conducted before most will accept the responsibility of consciously directed evolution. Even when more evidence is presented, there will be many who will still not believe in evolution. There is no doubt that it will take time. It will probably take decades for society to grapple with the social, philosophical and teleological implications of conscious evolution. It is a realisation that will grow in stages, and it will impact every aspect of our existence.

While the worlds of neo-Darwinian thought and Adam Smith's vision of a capitalist society were closely aligned so as to reinforce each other, the idea of conscious evolution and an unbridled capitalist world with new biotechnologies are not. Our resources are limited. We can improve farming efficiencies by using genetics to grow more vegetables and fruit in depleted soils full of heavy metals. We can use genetics to make chickens and pigs that grow faster and produce more protein. We can improve our farming efficiency by feeding animals on the recycled remains of other animal and bird carcasses, and keeping them in a smaller space. However, we do not know where these pursuits will ultimately lead us. Nor do we know the impact that these choices will have on our global gene pool. These questions transcend scientific boundaries.

There is another even darker view of a pessimistic future reshaped by the spectre of directing our own evolutionary future. It is to create a new world order to reinforce our existing world using eugenics. Many in the twentieth century believed that Lamarckian thought provided a sinister thread of influence linked to eugenics, and Adolf Hitler's policy of extermination. The idea of eugenics remains a lightning rod for the non-acceptance of Lamarckian views to the present. It is therefore a subject that I cannot ignore.

The word 'eugenics' was first used in the 1880s by Francis Galtin, who was a cousin of Charles Darwin. He believed that hereditary factors governed our character traits and our physical features. He used this argument to encourage 'good marriages'. In such a society eugenics was based on the principles of individual choice and freedoms.

Charles Davenport started working on human evolution at the Carnegie Institution at Cold Spring Harbor in New York in the early 1900s. He believed that genes were responsible for things like alcoholism and feeblemindedness. He once declared that, "human matings should be placed upon the same high plane as that of horse breeding".

In the first part of the twentieth century, the idea of eugenics started to permeate government policies, and some harsh new laws were enacted. These included compulsory sterilisation programs and racial segregation laws. The arguments put forward by intellectuals such as Charles Davenport were championed by Theodore Roosevelt, Winston Churchill and Calvin Coolidge.

Even George Bernard Shaw and John Maynard Keynes supported the decree that genius and other 'good' human qualities tend to be inherited. What these eugenicists had in common was that they placed value on humans according to the qualities that they possessed. They based the idea of 'value' on their own image, and they worked to maintain the fabric of a society that had given them great privilege.

Following this era however, the topic of eugenics rapidly became highly emotive and resulted in the state sanctioned deaths of millions of people. When the arguments of the intellectual and political supporters of eugenics were used to justify Adolf Hitler's 'solutions', there were many who were seduced into allowing the world to enter its darkest hours. The historical records of Hitler's Eugenic Sterilization Laws ordering compulsory sterilization of all those suffering from blindness, schizophrenia, or a wide range of hereditary afflictions and deformities should not be erased. Records of Hitler's extermination of millions of Jews and others in the war years should also never be forgotten. These serve as a reminder of the atrocities and social injustices that can be dictated by governments. Since then, post-World War II statutory laws have been introduced in western countries, and embodied by organisations like the United Nations (UN), to protect the rights of individuals.

It would be one of my greatest personal disappointments if these ideas of eugenics were resurrected as part of a plan to consciously direct our future evolution. If any of the concepts embodied in conscious evolution based on the new genetics were used to justify such decisions then my response would be this: The new genetics implies a responsibility for the individual. It implies responsibility for oneself and for the next generation. It implies the benefits of diversity, and that we should live in harmony.

There are also more possibilities to consider. Here I remain captive to my own inability—or unwillingness—to look beyond the immediate implications. Our children and grandchildren will have to create a future for HomoSap2.

TECHNOLOGY AND CONSCIOUS EVOLUTION

In the context of Lamarckian evolutionary learning logic, a pivotal point punctuating our evolution was when Homo Sapiens developed societies with inbuilt rewards for creativity. Whether the rewards were from the spoils of warfare, one to one combat or commerce, the result was a rapid revolution in new technologies.

We will continue to create new technologies to enrich and extend our lives, but only if we are rewarded for our efforts. While it is difficult for us to predict the type of new technologies created in the decades ahead, many of these will influence both the rate and direction of our evolution. Some will result in the direct manipulation of our heritable material, while others will influence our evolutionary direction by altering our behaviour.

We already know of many ways to manipulate genetic material. Our molecular systems are designed to accept new genetic information, or to delete or modify existing information. We are effectively 'growing our own' genes by engineering modified or enhanced versions of genes, and integrating these onto our genome. If genetic enhancement kits become available for domestic use, then we will have an even greater control over our destiny.

Other technologies are designed to provide additional sensory feedback from the environment. An electrocardiogram is really just pixels on a screen. But when a cardiologist looks at it, each pattern matches a conceptual image that is associated with a lot of other information stored in her head. Sometimes variations in the patterns or words occur that makes the cardiologist have to expand the range of conceptual images and their associations with other information. Thus, she augments her memory by inputting a variety of symbolic words and images derived from technological feedback. The feedback enhances the amount and type of information collected by the cardiologist, and it increases the speed and the efficiency with which she can make a medical diagnosis.

We are also reverse engineering our brain to understand how it works, and how it rewires itself so that we can improve our communication, memory and cognition. Perhaps we will develop new human interface technologies to master brain to brain communication, or add a nanobot surveillance system to our bloodstream for monitoring gene expression profiles. We are in the early stages of perfecting neural implants. In the future maybe we will be able to inject new RNAs into our bloodstream that represent a set of virtual experiences to enhance learning, quickly acquire new knowledge, or to develop new instincts. We might even be able to use the RNAs from a donor to acquire their knowledge in the form of past memories. Maybe a psychogenetic therapist will be able to synthesise and inject new genes into his patients.

There is no other technology that truly reflects our biological nature more than the internet. It has developed organically. It has a central nervous system, backup systems, memory and antiviral software to protect it against a wide range of possible malicious code attacks. It has evolved as a natural extension of our lives and the personal computer. In the future we may develop 'Google genes' that act to reinforce our online search behaviour using far more information than our current online footprints leave behind. As we build up our individual web search patterns, we leave behind a dynamic record of our behaviour. This information could be used as the source code for creating Google genes to improve our online search experience as the online environment becomes more enriched. That is, digital Google genes could be created as a reflection of our virtual lives using the same evolutionary learning logic that biological systems use. For example, when we go on holidays, our work Google search gene can be turned 'off', and the holiday Google gene switched 'on'—or expressed—using the same strategies and code efficiencies as that used by our biological genes. Triggers to activate one gene or the other would need to be set. The type of news stories highlighted, the screen display mode for different sites, the payment options offered for online purchasing, and the amount of assistance they may need can all be personalised by creating virtual genes that reflect our interactions with the virtual worlds we enter. Later, the Google gene family could be extended to include genes to build an avatar of ourselves. Genes for voice and character might be added to enhance text to voice or voice to text messaging. A lifelike avatar, complete with genes for finger prints and retinal scan data might be used to enhance the capabilities of online three-dimensional immersive technologies. As our lives continue to merge with technology, our virtual Google genes could become an important extension of our physical existence. At the same time, these interactions will help to direct our evolutionary future.

There are also new technologies that merge with biology. In the laboratory, we are able to work at a scale that is a million times smaller than a pinhead. This is nanotechnology. In this world the science of physics becomes governed by novel physical, micro-fluidic and quantum laws setting new challenges and limits on what can be created. At this scale, we are only just beginning to understand some of the tricks and forces that biological systems use to produce the outcomes and options available to them.

Some of the new nanotechnologies are inspired by nature. One example is research to design nano-fabrics that act like the skin of a chameleon. In a chameleon, the images the eye sees are translated into tiny electrical impulses that are sent to the brain. As a response, the brain sends tiny impulses along the central nervous system in the skin, and the skin cells each respond to the impulses by altering their colour. This is similar to mounting a sweating reaction in the skin or palms in response to a sudden shock. The nano-fabric is implanted with tiny image sensors that send electrical impulses to different parts of the fabric so that the colour in each tiny nano-cell unit is altered dynamically to reflect the environment. Once it is developed, a soldier moving through a hostile environment in such a fabric could become almost invisible to observers at will. Unsurprisingly, much of this research is classified.

An example of a nano-biotechnology development is provided by MIT scientist Angela Belcher. Angela has produced a virus that coats itself with semiconducting material and forms a bridge between two tiny electrodes.[111] The 'nano-wire' virus is intended as a precursor to a microbe she's developing that self-assembles into a transistor. In this case, we are starting to use biological DNA machinery to do our own engineering at the nanoscale. We can do this in our bodies, in our environment and fundamentally alter the role of a whole range of viruses, and bacteria. Just as we have learnt to domesticate animals to do some of our work, we might co-opt bacteria and viruses for constructing tiny nanoscale machines to build materials, new computing devices or medical cures.

While there are other new technologies that will influence the direction and rate of our evolution, the key point is that each of these new technologies is conceived and developed using the creativity arising from our consciousness. Each new stage in our future evolution will depend on our creativity. It is the key to unlocking the door to our own consciously directed evolution. In this sense, our computer systems, the internet, biotechnologies, and nanotechnologies are all just as much the result of our natural evolutionary processes as our physical form. They have all been developed as a natural extension of who we are, and how we are designed to function at the molecular level.

GENETIC DISCRIMINATION

Another way for societies to play a role in conscious evolution is through discrimination based on genetic testing. In the future, it is likely that testing for genetic and epigenetic signatures will reveal a lot more about our lifestyles and vulnerabilities than we can possibly know at present. We will probably soon be required to provide a blood sample for DNA profiling by life insurance companies, or to apply for some professional occupations. In the future, a five minute genetic test using data matching could be used to indicate what infections we have, or if we have diabetes. Genetic profiling will also be able to indicate our general fitness level, what cancers we are likely to develop, whether or not we are suffering from severe depression, and a whole range of other information about us, and our vulnerabilities.

Already there is evidence to suggest that genetic discrimination is occurring. In most countries, insurance companies are allowed to ask the customers if they have had any genetic screening. However, in countries like the U.K. there is a moratorium on insurance companies being able to use genetic test results until 2011.

In Australia it was reported by *New Scientist*[112] that insurers and employers are already discriminating against people on the basis of genetic data. In a survey of one thousand people who have taken a predictive genetic test, it was found that around one in twelve believed that they have been disadvantaged as a consequence. In one case, a woman tested positive to a variant of the BRCA1 gene that puts her at a higher risk of developing breast cancer. Her insurance company allegedly denied her life insurance cover for all forms of cancer.

The p53 gene is just one of many genes that might be used in future for routine profiling by life insurance companies. It is a single gene that can reveal a lot about us. In the late 1970s the p53 gene was accidentally discovered and identified as playing a key role in the development of cancer. By 2002, we knew that the p53 gene played a crucial role in protecting us from cancer, but little was known about how it functioned. We now know that the p53 gene acts as a tumour suppressor by detecting when the DNA of a cell is damaged. Once detected, it either activates a cell suicide program (*apoptosis*), or it prevents the cell from reproducing. The p53 gene becomes concentrated in cells that are damaged by excessive ultraviolet light exposure or radiation. While all cancers are caused by faulty genes, a mutant version of the p53 is the single most common genetic factor associated with all cancers.

In lung cancer cells of smokers, there is a specific 'hot spot' mutation on the p53 gene. In other words, your cigarettes leave a specific genetic identifier behind in your genes. In an age of pervasive genomic testing, you won't be able to lie to the insurance company about your smoking habits. They'll know. If your non-smoking children inherit your p53 gene signature associated with smoking, they may have to pay a higher life insurance premium because of the associated increased risk of some other forms of cancer. Exposure to aflatoxin, a poison secreted by a fungus and exposure to an excessive amount of ultraviolet light also leave identifiable signatures on the p53 gene.

Damage to the p53 gene may also be related to an increased risk of some other types of cancer. In 2001, studies of the p53 gene were published showing that smokers diagnosed with a malignant melanoma have a poorer prognosis because they have an increased risk of metastases.[113]

There are also other medical implications associated with the p53 gene. We know that there are a small percentage of people who have a damaged p53 gene.

Most inherit it, and they are extremely vulnerable to cancer from birth. The p53 can also become disabled by other means. In the case of the sexually transmitted human papilloma virus (HPV), it deactivates the p53 gene so that all of the HPV cancer cells do not have the tumour-suppressing ability conferred by the p53 gene.

In 2002, some scientists working with genetically engineered mice accidentally found that the p53 gene was also related to ageing.[114] They knocked the p53 gene out of some mice. When they bred from the mice it was observed that they developed tumours at a very early age as expected. They bred another group of mice with a modified p53 gene. The modified mice and their offspring proved to be well protected from cancer. But what surprised the scientists was that within a few months, they looked like very old mice. Their fur became grey and ruffled, they developed hunchback spines and they lived only about two-thirds as long as expected. While we know that our chances of getting cancer increase with age, no one expected that a single gene may form some link between the two. A key question it raises is, can cancer treatments designed to stimulate the p53 gene to protect us from cancer also accelerate the ageing process, leading to age-related ailments like bone loss or dementia?

Despite being able to tell a lot about ourselves from the p53 gene profile, these findings reveal that we still do not understand the other relationships that exist between the p53 and other genes.

As our ability to rapidly sequence larger amounts of genetic data improves, we are uncovering a greater number of genetic links to disease. Using tiny disposable biochips smaller than a USB thumb drive, we will soon be able to rapidly identify a large number of links between our genes and our lifestyles. It may not be long before a new generation of direct genetic sequencing technologies able to screen a whole genome within a few seconds are put into use to identify the risks associated with over four thousand heritable diseases that we have so far identified. Several laboratories are working on nanoscale fast sequencing technologies that could potentially directly sequence millions of bases in a few seconds.

One group of nanoscale fast sequencing technologies that seem promising uses a 'nanopore' similar to the ones used to allow nucleic acids to pass across cell membranes. The nanopore system detects tiny electric field differences arising from the different molecular structures of bases A, G, C, and T (or U) as they sequentially pass through a nanopore. As the test DNA (or RNA) strands pass through the nanopore, tiny sensors will be able to use the small variations in ionic current to differentiate between the different molecular structures of each base. The nucleic acid sequence will then be analysed as it squeezes through the nanopore. By successfully developing technologies like this, a whole genome could be sequenced and analysed within a few seconds. However, I should stress, that while scientists are working on these concepts in several test laboratories, the technology does not yet exist. We still have a lot to learn about working at this scale in the laboratory.

Once low cost, fast sequencing technologies become available, non-invasive whole genome screening will become routinely used by health centres. But this raises some thorny ethical and social issues. Do we abort a female foetus with a ninety five percent chance of contracting breast cancer before she is twenty five years old? Should parents be made aware of the risk profiles associated with certain diseases for their foetus? What about tendencies to exhibit aggressive behaviours, addictive behaviours such as alcoholism, or obesity on the basis of genetic screening of a foetus?

There is already evidence that the confronting prospect of designer babies could go well beyond selecting the sex of a foetus and screening it for a range of congenital diseases. While most traits are polygenic, genomic screening could soon extend to choosing a designer baby with the right height, IQ, eye colour, weight, looks or personality traits. The rich, who can pay for the expensive technologies and specialists required, will probably tend to choose offspring with all the 'good' genes. Governments could even the playing field by subsidizing some gene therapies, and regulating to prevent or limit the use of others. It will not be easy to determine which services will be funded. It will involve politics, budgetary considerations, company interests, national interests, religion and, of course it will draw on our values.

Another area where new screening technologies are likely to play a pivotal role in conscious evolution is in the use of brain scanning technologies. It is already becoming possible to "read people's minds" by looking at scans using magnetic resonance imaging. Tom Mitchell who has led some early experiments at the Machine Learning Department at the Carnegie Mellon University in Pittsburgh was surprised to discover just how similar the pattern of observed brain activity was among volunteers involved in his experiments.[115] The ability to see what people are thinking is limited to identifying functional image differences as people are thinking about one specific word such as 'aeroplane' rather than another, such as 'celery'. These early experiments provide us with new avenues to learn more about how we store information and our ability to characterise links between our thoughts and molecular data.

While we cannot be certain about how new gene technologies will be used, or how we will regulate their potential for discrimination, they are set to revolutionise medicine and play an important role in creating a future for HomoSap2.

SO WHERE TO FROM HERE?

To this point we have fine-tuned ourselves to compete in a world where only the fittest survive. We have built up our political systems, and economic structures based on this belief. We have fought wars to defend territories and beliefs. The animal kingdom has survived this way as well. But how will a world that is conscious of its ability to direct our future evolution, help us to survive the next challenges? How will it change how we think?

A unifying implication is that it will result in a new way of thinking about survival as a species and our future. Based on this reasoning, our brains then are ready and able to expand the meaning we attach to ourselves and our ideas on what our future might hold. Consciously directed evolution provides the basis for an expanding domain of life, our understanding of consciousness, and our relationship with new technologies. While some of the technologies and possibilities for discrimination considered in this chapter remain in the realms of science fiction for the present, how we use them will play a pivotal role in moulding our future evolutionary direction.

What values will guide us through the challenges of accepting the idea of genetic responsibility? To meet the challenges, perhaps HomoSap2 will require a larger brain after all! In the meantime, I will go sailing and continue to contemplate what it all means.

* * *

219219

219

ACKNOWLEDGEMENTS

I owe a special thanks to the many people who have helped me to deliver the messages of the *Soma* to the famed general reader. In particular, I thank Julian Steele for his research assistance, and for creating a citation index. I also thank my editor Hazel Flynn for her professionalism, her insight and her rigorous editorial assistance. She has encouraged and helped me to communicate with the general reader. I thank Bronson Boyd for his suggested 'one liners' that have helped to make the text more colourful. I am indebted to Dr Ted Steele (immunologist, co-author of *Lamarck's Signature* and my husband), Michael Denton (human geneticist, author of *Evolution: A Theory in Crisis* and *Nature's Destiny*) and Geoff Pain (biochemist) for their scientific input, and some sound editorial advice. I acknowledge Ted Steele, Eric Nestler and Adrian Gibbs for permission to use their quotations. I am also indebted to Professor Roger Dawkins and the other scientists at the CY O'Connor Education Research and Development Employment (ERADE) Village Foundation. I consider it both an honour and a privilege to be a part of the multidisciplinary CYO research team. I am discovering a rich bounty of new ideas on evolutionary thought at our regular 4:00 pm seminars where debates frequently extend well into the night at the CYO Village Pub. A special thank you to Dianne, Kelly and Janelle from the Tin Shed cafe in Thirroul: your endless supply of great coffee has provided me with sustenance when I have needed it most. Last, I thank Mary Sokopf and Anne Gregory for their unconditional friendship and support. Without your help, the journey would not have been possible.

APPENDIX

Figure 1.   Diagram showing the pangenomic model consisting of the genome and the epigenome. Arrows show the main information pathways operating at each level within the soma.

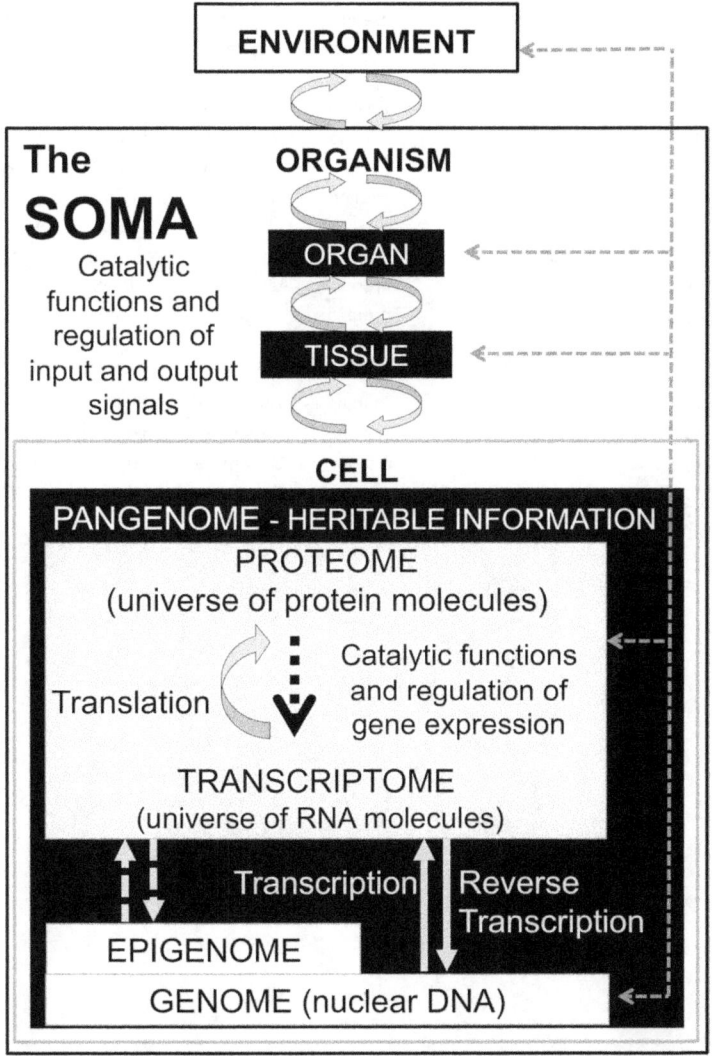

Table 1.   Table showing the twenty standard amino acids grouped according to their triplet code of bases (codon).

Amino Acid Formation

First base	Second base				Third base
	U	C	A	G	
U	Phenylalanine	Serine	Tyrosine	Cysteine	U
U	Phenylalanine	Serine	Tyrosine	Cysteine	C
U	Leucine	Serine	Stop	Stop	A
U	Leucine	Serine	Stop	Tryptophan	G
C	Leucine	Proline	Histidine	Arginine	U
C	Leucine	Proline	Histidine	Arginine	C
C	Leucine	Proline	Glutamine	Arginine	A
C	Leucine	Proline	Glutamine	Arginine	G
A	Isoleucine	Threonine	Asparagine	Serine	U
A	Isoleucine	Threonine	Asparagine	Serine	C
A	Isoleucine	Threonine	Lys	Arginine	A
A	Methionine or Start	Threonine	Lys	Arginine	G
G	Valine	Alanine	Aspartate	Glycine	U
G	Valine	Alanine	Aspartate	Glycine	C
G	Valine	Alanine	Glutamate	Glycine	A
G	Valine	Alanine	Glutamate	Glycine	G

Note: In DNA, U = T

GLOSSARY

amino acid The basic chemical building blocks of proteins. (Also see Table 1 in the Appendix.)

antibody A protein structure produced by B cells in response to foreign antigens (bacterial cells, virus particles and their toxic products).

antigen Any molecular structure capable of specific binding to an antibody or a T cell receptor.

base Refers to A (adenine), T (thymine), G (guanine), C (cytosine) in DNA, and the base U (uracil) which replaces T in RNA.

B cell A white blood cell produced in the bone marrow that matures into antibody producing B cells that can be bound and 'selected' by the antigen leading to activation into antibody secretion or cell proliferation.

cadherin genes The family of genes required for the formation of the protein structures responsible for cell-cell adhesion and communication (e.g. in nerves and muscle tissue).

chromosome A long double stranded DNA molecule complexed with protein and containing many genes.

codon Sequence of three adjoining bases specifying a particular amino acid. (Also see Appendix, Table 1.)

DNA (deoxyribonucleic acid) A nucleic acid composed of two chains of nucleotides A (adenine), G (guanine), C (cytosine) and T (thymine). The two chains are joined together by a ladder of hydrogen bonds, and shaped into a double helix.

endocrine system The network of endocrine glands in an organism that is responsible for secreting a range of hormones directly into the bloodstream.

enzyme A protein that acts as a catalyst for a biochemical reaction.

epigenome The total complement of modifications to the surface of a DNA structure that alters the expression of a gene, without altering the DNA sequence.

gene A segment of DNA that encodes the information required to produce a single functional biological product (protein or RNA).

genome All of the genes (DNA) contained in a single set of chromosomes.

germline All genes (DNA) encoded in the germ cells (or sex cells).

horizontal gene transfer Any process which results in an organism integrating genetic material from another organism.

hypermutation A process that involves an abnormally high rate of mutation in the variable regions of immune system genes. It may also occur in the variable regions of some other gene families (e.g. in some cadherin genes).

Lamarckism Today, the concept of 'Lamarckism' is understood as changes in an individual that are acquired during its lifetime and are inherited by its offspring.

lymphocytes The white cells in blood that play an important role in defending the body against disease. The main types of lymphocytes are the B cells and T cells.

mRNA Messenger RNA is the group of RNAs that are used to make proteins (i.e. for protein translation).

major histocompatibility (MHC) complex In the immune system, the MHC genes play a key role in enabling T cells to distinguish between 'self' and cells containing foreign viruses. Some MHC genes have also evolved to assist the whole organism to use its olfactory system to distinguish between different MHC types. Many of the odorant receptor genes are in the MHC (histocompatability complex) region of our genome.

methylation A chemical reaction that results in the addition of a methyl group (-CH3) to a molecule.

morphogenetic Physical changes in form and structure.

nucleic acid The chemical name for all DNA and RNA molecules.

nucleotides The basic chemical structures, which, with the bases (A, T, C, G or U) link together to form strands of DNA or RNA.

olfactory system The sensory system used for olfaction (involving the sense of smell).

ova Unfertilized egg cell.

pangenome The heritable somatic elements contained within the epigenome and genome. The term has been coined for use in this text to provide a single word description of all of the somatically derived heritable factors that can be transferred to the next generation. It is derived from the term 'pangenesis' which is an ancient theory that hereditary characteristics were carried and transmitted to progeny by 'gemmules' derived from somatic cells.

pathogen Virus or bacterium causing disease.

phenotype Observable physical and biochemical characteristics of an organism.

point mutation A change to a single base in a DNA or RNA sequence.

prion Small protein molecule that may be able to direct its own replication. The idea that they might be also able to transmit heritable information directly from one protein molecule to another is relatively new.

protein Long polymer of amino acids folded into a complex three dimensional arrangement.

reverse transcription The synthesis of a complimentary copy of DNA from an RNA template.

reverse translation A speculative process involving the transfer of information encoded in the form of protein into RNA or DNA.

retrovirus RNA virus that makes a DNA copy of itself by the process of reverse transcription.

replication The process of duplicating another copy of DNA or RNA.

RNA (ribonucleic acid) Single stranded (usually) copies of a DNA sequence. In RNA a U (uracil) is functionally equivalent to a T (thymine) in DNA.

somatic Of the body (i.e. not a part of the germline).

stem cell One of the bodies master cells that have the ability to differentiate into any one of the bodies cell types.

T cell Thymus derived white blood cells that express antibody-like surface molecules. While antibodies are required to fight a bacterial infection, the T cells are required to fight viral infections.

telomere The tip of a chromosome that contains a number of repeats of the same base sequence.

transcription The process of synthesising RNA from a DNA template.

transgenic An organism whose genome has been technically altered by the introduction of new genetic material.

translation The process of synthesising protein from an RNA template.

virus A single RNA or DNA molecule that is surrounded by a protective protein coat.

INDEX

A

Ader and Cohn, 142-3
Affinity maturation, 99
AIDS, 80, 102-4
Alloxan
    and induced diabetes, 90
Alzheimer's
    disease, 35,66
Antibiotic resistance, 40-2
Antibodies, 90-102
Aristotle, 2

B

B cell, 99-102
Bacteria,
    acquired inheritance, 38-44
Baldwin
    James Mark, 180
Balfour
    Robert, 23
Baltimore
    David, 78-9
Belcher
    Angela, 210-11

Benois
    Jacques, 29-30
Blackburn
    Elizabeth, 81-2
Bodnar
    Andrea, 82
Bone remodelling, 135-6
Brain
    human, 189-93
Brink
    Alexander R., 116-7
Buck and Axel, 172-3

C

Cadherin and antibody genes
    parallels between, 140-3
Cadherin genes, 137-43
Cairns
    John, 42-3
Canary
    reproduction, 121-4
Cell factory, 71-7
Central Dogma, 78
Chameleon, 210

Chardin
   Teilhard de, 197
Churchill
   Sir Winston, 205-6
Cicada's internal clock, 122-4
Cichlid fish, 151
Cocaine
   addictive behaviour, 137
Cohen, Boyer and Berg, 79
Colorectal cancer, 119
Connolly
   Dorothy, 26-7
Conscious evolution
   technology, 207-11
   the idea of, 197
Coolidge
   Calvin, 205-6
Copernicus
   Nicholas, 159-61
Crick
   Francis, 62-3. 78, 127
Cuvier
   Georges, 6
Cystic fibrosis, 74

D

Darwin
   Charles, 8-13, 23-5
Davenport
   Charles, 205-6
Dawkins
   Richard, 163

Death Cap mushroom, 65-6
Diabetes, 90, 115-6
Diet
   acquired inheritance, 113-6
Dolly
   the sheep, 111
Ducks, Khaki Campbell
   inheritance effects, 29-31
Dyson
   Freeman, 34-5

E

E. Coli bacteria, 74
Einstein
   Albert, 197
Eldredge and Gould, 150-3
Elefteriou
   Florent, 136
ENCODE project, 61-2
Ephestia, 30
Epigenetics
   behaviour, 128-34
   betel nut and diabetes, 115-6
   definition, 108
   diet, 113-6
   disease, 118-20
   education, 132-4
   genome-wide response,
      121-4
   honey bee, 114-5
   leaving footprints, 107-8, 134
   marbled crayfish, 107-13

Eugenics, 205-7
Evolution
    the emerging paradigm, 164-6
Evolutionary learning
    immune system, 170-1
    language, 175-7
    logic, 167-9
    odorant receptor genes, 171-3
    social change, 182-7

F

Fear, 132
Feig
    Larry, 132-3
Flu
    virus infection, 95-6
Fogarty
    Patrick, 31-3
Fraga
    Maria, 111-2
Fruit flies, 19
Functional shift, 135

G

Galileo
    Galilei, 159-60
Galton
    Francis, 205
Gene gun technologies, 26
Genetic crop engineering, 26

Genetic discrimination, 211-7
Genetic memory, 143-7
Genetic responsibility, 200-2
Genetic tharapy, 35-8
Gibbs
    Adrian, 94
Goldner and Spergel, 90
Google genes, 209-10
Gorczynski
    Reg, 25-6, 91-2
Gould
    Stephen Jay, 150-3
Grant
    Robert, 7

H

Heraclitus
    of Ephesus, 167
Hershey and Chase, 55
Hippocrates of Cos II, 1,11
Hitler
    Adolf, 205-7
HomoSap2, 205-7
Hooker
    Joseph, 9
Horizontal gene transfer, 21
Hoyle
    Fred, 157-8
Human Genome Project, 56-8
Human papilloma virus
    (HPV), 81

Huxley
  Aldous, 'Brave New World', 147-8
  Julian, 197
  Professor Thomas, 13
Hypermutation
  in B cells, 99-100

I

Identical twins, 112-3
Immune system
  acquired inheritance, 89-104
Imprinting
  genomic, 120-1
Instinct, 80, 185-6
Intelligence
  measures and social context,
    181-2

J

Jung
  Carl, 185-6
Jungian archetypes, 185

K

Kammerer
  Paul, 17
Keynes
  John Maynard, 206
Kit gene, 117
Koala
  AIDS (KoRV), 102-3

Kolliker
  Albrecht von, 49
Kuhn
  Thomas, 161
Kuru
  disease, 66

L

Lamarck
  Jean-Baptiste de, 3-6, 12-3,
    162-5
Lamarck's Signature, x, 94
Lubbock
  John, 13
Lyell
  Sir Charles, 8, 11, 13
Lysenko
  Trofim, 18-9

M

MacLeod and McCarty, 39
Macrocephali people, 1
Mad Cow disease, 66
Major histocompatibility
    complex (MHC),
    173-5
Malaria, 96
Marbled crayfish, 109-12
Marchetti and Wryobek, 52-3
Mattick
  John, 86

McClintock
  Barbara, 45
McDougall
  William, 17-8
Medawar
  Peter, 182
Melanoma, 119
Mendel
  Gregor, 13-4
Mendelian genetics, 13
Miller and Sweatt, 128
Mitchell
  Tom, 216-7
Moore
  Thomas, 203-4
Morton
  Lord, Arabian chestnut mare, 24
Muller
  Fritz, 13

N

Nano-pore, 215
Nestler
  Eric, 130

O

Odorant receptor genes, 171-3
Okamoto, K, 90
Origin of Species (The), 9, 12
Ova
  or ovary, 49

Overthrowing an old scientific
  paradigm, 158-62
Oxytricha, 84

P

p53 gene, 212-4
Pangenesis,
  Theory of, 1, 3, 11
Parkinson's disease, 66, 74
Pavlov
  Ivan, 16-7
Pekin ducks, 29-30
Philosophie Zoologique
  by Lamarck, 4-8, 12
Placebo effect, 142-3
Plato
  Greek philosopher, 149
Plato's Cave, 149
Pope
  John Paul II, 159-60
Popper
  Karl, 182-3
Prenant
  Marcel, 19
Prions, 43
Punctuated Evolution, 150-3

R

Rassoulzadegan
  Minoo, 117-8
Recombinant DNA, 31

Retrovirus, 76
Reverse translation, 83-4
Reverse transcription, 77-83, 114
RNAs, 68-71
Roman Catholic Church
    Copernicus, 160
Roosevelt
    Theodore, 205-6
Rostrand
    Jean, 30
Royal Society of London, 3

S

Salmonella bacterium, 65
Schizophrenia, 136-7
Shaw
    George Bernard, 206
Silkworm larva
    translation, 65
Sire effect, 22-9
Smallpox, 95
Smith
    Adam, 204
Smoking and DNA damage, 52-3
Snow
    Allison, 46
Sobey
    Bill, 26-7
Social evolution
    as a Lamarckian phenomenon,
        182-7
Socrates, 1

Soma
    a drink, 147-8
    whole body, 200-2
SOS proteins, 75
Spadafora
    Corrado, 50-1
Spencer
    Herbert, 23-4
Sperm
    chemical damage, 52-3
    derivation, 49
    surface as a vector, 50-2
Steele
    Ted, 91-5
Strzelecki
    Count, 24
Suicide
    genetic change in brain, 130
Superbugs, 44-7
Superweeds, 44-7
Synaptic plasticity, 134-7
Szyf
    Moshe, 130

T

T cell, 80, 97-9, 102, 143, 174
Telomeres, 82
Temin
    Howard, 77-8, 91
Thalassemia, 96
Transcription, 60-2
Transgenic animals, 31-5

Translation, 62-8, 73
Twins
    studies of, 136

V

Vatican Science Panel, 160
Venter
    Craig, 57
Vogt
    Gunter, 110

W

Waddington
    Conrad, 108

Wallace
    Alfred, 8, 13
Watson
    James, 56-8
Weaver
    Ian, 131-2
Weismann
    August, 14-5, 25
Weismann's barrier, 91

Y

Yeast
    Sup35 protein, 43

REFERENCES AND NOTES

1   Steele, Edward J., Robyn A. Lindley and Robert V. Blanden (1998). Lamarck's Signature: How Retrogenes Are Changing Darwin's Natural Selection Paradigm. Perseus Books, Reading, Massachusetts.

2   Landman, Otto (1993). Inheritance of acquired characteristics revisited. BioScience 43(10) 696-705. Some of these are described in: The Anthropological Treatises of Blumenback and Hunter (1865) pp. 241-243 (books.google.com). There are many other reports, and some prints of people adopting the same practice that resulted in infants being born with elongated heads.

3   Some of these are described in: The Anthropological Treatises of Blumenback and Hunter (1865) pp. 241-243 (books.google.com).

4   As an example, Anaxagoras and the anatomists.

5   Lamarck, J.B. (1809). Zoological Philosophy: An exposition with regard to the natural history of animals.

6   A clear and rather sympathetic exposition of Lamarck's ideas is found in: Gould, Stephen Jay (2002). The structure of evolutionary theory. The Beknap Press, of Harvard University Press.

7   Grauer, Dan, Manola Gouy and David Wool (2009). In Retrospect: Lamarck's treatise at 200. Nature 460 688-9.

8   For a critical discussion of the contributions made by Charles Darwin, and the influence of his ideas refer to: Bowler, Peter J. (1990). Charles Darwin: The Man and His Influence. Cambridge University Press.

9   A far more detailed comparison and analysis of the works of Lamarck and Darwin is provided by Stephen J. Gould in: Gould, Stephen J. (2002). The Structure of Evolutionary Theory. Harvard: Belknap Harvard.

10 Windholz, George and P.A. Lamal (1991). Pavlov's view of the inheritance of acquired characteristics as it relates to theses concerning scientific change. Synthese 88 97-111.

11 While the debate has continued over his innocence, recently published analysis suggests that his experiments may have been the first demonstration of epigenetic change induced by the environment. See: Vargas. Alexander O. (2009). Did Paul Kammerer Discover Epigenetic Inheritance? (2009). A Modern Look at the Controversial Midwife Toad Experiments. Journal of Experimental Zoology (Mol. Dev. Evol.) 312B

12 Quoted in: McDougall, William (1927). An Experiment for the Testing of the Hypothesis of Lamarck. British Journal Of Psychology, 17(4), p. 271.

13 McDougall, William (1930). Second Report on a Lamarckian Experiment. British Journal Of Psychology, 20(3), pp. 201-18.

14 The sire effect is also variously referred to by animal breeders as 'throwing back', or by its scientific name 'telegony'.

15 A wide range of examples involving both animals and humans are described in: Gould, George M., M.D. Walter and Walter L. Pyle (2002). Anomalies and Curiosities of Medicine: Paternal Impressions, Telegony Part 2.

16 Gould, G.M. and W.L. Pyle (1896). Anomalies and Curiosities in Medicine. Philadelphia. Reprinted online by Google, Plain Label Books, 2003 p. 63.

17 Dunglison, Robley (1856). Human Physiology. Blanchard and Lea, Philadelphia, Eight Ed., Vol II. Digitized by Google, p. 474.

18 Darwin, Charles (1868). The Variation of Plants and Animals under Domestication. Orange Judd and Co., Volumes I and II.

19 Darwin, Charles (1859). The Origin of Species. Chapter V, Laws of Variation.

20 Darwin, Charles (1868) Variation in Plants and Animals Under Domestication. Volume I.

21 Gorczynski, R.M. et al (1983). A possible maternal effect in the hyporesponsiveness to specific alloantigens in offspring born to neonatally tolerant fathers. The Journal of Immunology 131 1115-20.

22 Sobey, W.R. and Dorothy Conolly (1986). Myxomatosis: Nongenetic Aspects of Resistance to Myxomatosis in Rabbits, Oryctogagus Cuniculus. Australian Wildlife Research, CSIRO 13 177-87.

23 Watson, J. Gary et al (1983). Reproduction in Mice: The Fate of Sperm Cells Not Involved in Fertilization. Gamete Research 7 75-84.

24 Benoit, J., LeRoy P., Vendrely R. And Vendrely C. (1960). Experiments on Pekin Ducks treated with DNA from Khaki Campbell Ducks. Transactions of the New York Academy of Science 22 494-503.

25 Science: At the Mainsprings, Time Magazine Monday, 5th August 1957.

26 Nawa, Saburo and Masa-Aki Yamada (1968). Hereditary change in Ephestia after treatment with DNA. Genetics 8 573-84.

27 These are known as 'retroviral vectors'.

28   Fogarty, Patrick (2002). Optimizing the production
     of animal models for target and lead validation.
     Journal of Drug Discovery Today (previously called
     'Targets), 1(3) 109-116.
29   The 'transposon' gene expressed by the vector allows
     the normal integration of the DNA taken up by
     somatic and germ cells i.e. you rely on the presence
     of the cell's natural molecular machinery for the
     integration system to be activated.
30   Dyson, Freeman (2006). Make me a hipporoo.
     NewScientist Weekly, 11 February 2006, pp.36-9.
31   A vaccine is a preparation of microorganisms or
     their antigenic components designed to induce
     immunity against a specific pathogen (either viral
     or bacterial), without causing disease in the host.
32   Avery, O.T. et al (1944). Studies on the chemical
     nature of the substance inducing transformation of
     pneumococcal types: Induction of transformation
     by a deoxyribonucleic acid fraction isolated from
     pneumococcus. Journal of Experimental Medicine
     79(2) 137-58.
33   Adam, Mike et al (2008). Epigenetic Inheritance
     Based Evolution of Antibiotic Resistance in Bacteria.
     BMC Evolutionary Biology, BioMed Central 8(52).
34   Cairns, J. et al (1988). The Origin of Mutants. Nature
     335 142-5.
35   The special proteins referred to are 'prions'.
     Although the prion model was originally applied
     to infectious diseases, it is now studied as an agent
     for some acquired inheritance effects. The idea
     that proteins like prions can transmit information
     directly from one protein molecule to another
     protein molecule is relatively new. For a review refer
     to, Chernoff, Yury O. (2000). Mutation processes

at the protein level: Is Lamarck back? Reviews in Mutation Research 488 39-64.

[36] For two excellent reviews of the field that point out the Lamarckian nature of the phenomenon described refer to: Rando, Oliver J. and Kevin J. Verstrepen (2007). Timescales of Genetic and Epigenetic Inheritance. Leading Edge Review, Cell 128 655-668; and, Chernoff, Op. Cit. pp. 39-64.

[37] McClintock, B. (1978). Mechanisms that rapidly reorganize the genome. Stadler Symposium 10, pp. 25-48.

[38] For information on a range of publications by the Institute of Science in Society (ISIS) in London, refer to their web site at: www.i-sis.org.uk. ISIS has produced many important publications warning of the social and ethical consequences of not understanding the fluid nature of our genome. Information on many of the unintended consequences of releasing genetically modified crops into the environment can be found in a comprehensive dossier of referenced articles presented to the European Parliament, GM Science Exposed: Hazards Ignored, Fraud, Regulatory Sham, Violation of Farmer's Rights, June 12, 2007.

[39] These are called embryonic stem (ES) cells. It is also interesting to note that donor-derived oocytes have been observed in adult female mice following peripheral blood transplantation. Further information about some of the more interesting aspects of oocyte and sperm production refer to: Johnson, Joshua et al (2005). Oocyte Generation in Adult Mammalian Ovaries by Putative Germ Cells in Bone Marrow and Peripheral Blood. Cell, 122 303-15; and, Geijsen et al (2004). Derivation of embryonic

germ cells and male gametes from embryonic stem cells. Nature, 427 148-54.

[40] Spadafora, Corrado (2007). Sperm-mediated Gene Transfer: Mechanisms and Implications. Spermatology, 65 459-67.

[41] Most of the foreign sequences are replicated as 'extrachromosomal episomes' with mosaic tissue expression (i.e. not expressed in all tissues).

[42] Marchetti, Francesco and Andrew J. Wyrobek (2008). DNA repair decline during mouse spermiogenesis results in the accumulation of heritable DNA damage. DNA Repair 7 572-81.

[43] Wadman, Meredith (2008). James Watson's genome sequenced at high speed. Nature, News 425 788.

[44] For further background information on the history of our thinking about the gene and some contemporary and unorthodox views of what a gene is refer to: Evelyn Fox Keller called The Century of the Gene, published by Harvard University Press, 2000.

[45] One exception is polyproteins that are 'cut' by specific enzymes into many smaller polypeptides e.g. in neurotransmitters of the central nervous system.

[46] A number of agents (e.g. 'prions') and pathways have been implicated as being involved in the transmission of diseases involving protein misfolding.

[47] Temin, H.M. and S. Mizutani (1970). RNA-Dependent DNA Polymerase in Virions of Rous Sarcoma Virus. Nature, 226 1211-13. Baltimore, David (1970). Viral RNA-dependent DNA Polymerase. Nature, 226 1209-11.

[48] Storici, Francesca et al (2007). RNA-templated DNA repair. Nature Letters, 447 338-41.

[49] For a discussion on some of the other potential roles for telomerase refer to: Blackburn, Elizabeth H.

(2005). Shaggy mouse tales. Nature News & Views, 436 922-3.

[50] Lange, Titia de (1998). Telomeres and Senescence: Ending the Debate. Science, 279(5349) 334-5.

[51] While this is true, there remain some curious exceptions, and there is some evidence to suggest that in early evolutionary history some cells may have had the molecular machinery to support reverse translation. For further information on work by Nashimoto refer to: Nashimoto, M. (2001). The RNA/Protein Symmetry Hypothesis: Experimental Support for Reverse Translation of Primitive Proteins. Journal of Theoretical Biology 209 181-7.

[52] A team of evolutionary biologists working with Mariusz Nowacki at Princeton University, New Jersey has shown that this is the mechanism used in Oxytricha a ciliated protozoa, where the phenomenon has been well studied. This process is known to occur extensively across the genome in ciliates. In experiments reported by Nowacki and his colleagues, it was shown that the process is guided by parental RNA templates that are used to direct the DNA reassembly across generations. These templates appear to be critical in organizing the DNA rearrangements. They also have the potential to ensure that new epigenetic effects embodied in the RNA regulatory network reach the genome. In the RNA-mediated rearrangement of the genome, hundreds of thousands of different tiny fragments are scrambled. They are reassembled only after each fragment has had the opportunity for short sequence dependent comparison and editing to occur between the germline and somatic genomes.

It is not yet clear if DNA templates are also used as assembly guides to recreate a new macro-nuclear chromosome. However, it is evident that RNA templates and the RNA regulatory network play a very active role in ensuring that the DNA and the RNA network are both updated in the process. This work has provided the first evidence that the presence of a maternal RNA template that are effectively RNA-cached copies of DNA sequence from a previous generation, can guide genome-wide rearrangements in vivo. See: Nowacki, Mariuz et al (2007). RNA-Mediated Epigenetic Programming of a Genome-Rearrangement Pathway. Nature, 451 153-8.

[53] Mattick, John (2007). A New Paradigm for Developmental Biology. Journal of Experimental Biology, 210 1526-47.

[54] Okamoto, K. (1965). Apparent Transmittance of Factors to Offspring by Animals with Experimental Diabetes. In: On the Nature and Treatment of Diabetes, B.S. Liebel and G.A. Wrenshall (Eds). Excerpta Med. Amsterdam, pp. 627-31.

[55] Goldner, M.G. and G. Spergel (1972). On the Transmission of Alloxan Diabetes and Other Diabetogenic Influences. Advances in Metabolic Disorders, 6 57-72.

[56] A review of these and other early acquired inheritance experiments involving endocrine organs can be found in: Campbell, J.H. (1982). Autoevolution In: Perspectives on Evolution. R. Milkman (Ed.). Sinauer Associates, New York, pp. 190-201.

[57] Steele, E.J. (1979). Somatic Selection and Adaptive Evolution: On the Inheritance of Acquired Characters. Williams and Wallace, Toronto, 1979.

A 2nd edition was published by University of Chicago Press, Chicago in 1981.

58 Gorczynski, R.M. and E.J. Steele (1980). Inheritance of acquired immunological tolerance to foreign histocompatibility antigens in mice. Proc. Natl. Acad. Sci. USA 77(5) 2871-5.

59 A summary of some of the additional acquired inheritance experiments conducted by Ted Steele and his colleagues in the early 1980s can be found in: Steele, E.J., R.M. Gorczynski and J.W. Pollard (1984). Evolutionary Theory: Paths into the Future. Edited by J.W. Pollard, John Wiley and Sons Ltd., Chapter 9, 217-37.

60 Brent, L. et al (1981). Supposed Lamarckian inheritance of immunological tolerance. Nature, 290 508-13.

61 Steele, T. (1981). Lamarck and Immunity: A Conflict Resolved. New Scientist, May 7, 362-1.

62 Mullbacher, A. et al (1983). Induction of T Cell hyporesponsiveness to Bebaru in mice, and abnormalities in the immune response of progeny of hyporesponsive males. Australian Journal of Experimental Biology and Medical Science 61(2) 187-91.

63 Steele, E.J. (1988). Observations on Offspring of Mice Made Diabetic with Streptozocin. Diabetes, 37, pp. 1035-43.

64 The story of Ted Steele's battles to have his work on Lamarckian inheritance effects recognised is described in: Honeywell, Ross (2008). Lamarck's Evolution: Two Centuries of Genius and Jealousy. Pier 9, an imprint of Murdoch Books Australia.

65 Steele, Edward J., Robyn A. Lindley and Robert V. Blanden (1998). Lamarck's Signature: How Retrogenes Are Changing Darwin's Natural Selection Paradigm. Perseus Books, Reading, Massachusetts.

66 T cell receptors (TCRs) look at the major histocompatibility (MHC) proteins that are bound to a peptide digestion product of the virus. TCRs look at the distortion in the three-dimensional structure of the MHC protein structures.

67 Recent research has shed much light on the somatic hypermutation (SHM) process. It involves some direct mutations of the antibody gene sequence and error-prone repair of mutated DNA sequences which now seem to involve an mRNA template intermediary, or the involvement of RNA editing coupled to an error-prone reverse transcription step to incorporate the changes made in the RNA into the DNA. The first evidence that it involves an RNA intermediary was provided by us in 2006. See: Steele, E.J., Robyn A. Lindley, Jiayu Wen and Georg F. Weiller (2006). Computational analyses show A-to-G mutations correlate with nascent mRNA hairpins at somatic hypermutation hotspots. DNA Repair 5 1346-63. Just as the reviews for this manuscript were being finalised, we made a further important discovery: RNA intermediates appear to be involved in the genesis of somatic mutations in several non-lymphatic cancer genomes. This implies that whilst somatic hypermutation is a tightly regulated and beneficial process for B cells, aberrant mutations (or 'crises') in this complex mechanism in a range of somatic tissue types could result in cancer. This in turn suggests potential targets involved in SHM such as AID, Pol-eta, ADAR1 and modulators of the RNA Pol II transcription-coupled repair (TCR) apparatus should be considered as possible new drug targets in the development of future cancer therapies. See:

Steele, Edward J. and Robyn A. Lindley (2010). Somatic mutation patterns in non-lymphoid cancers resemble the strand biased somatic hypermutation spectra of antibody genes. Letter to the Editor, DNA Repair 9 600-3. Copies available on request.

[68] Rachael E. Tarlinton et al (2006). Retroviral invasion of the koala genome. Nature 442 79-81.

[69] The epigenetic machinery differentially controls the expression of a whole range of genes that enables each type of cell to perform different functions. While the DNA sequence may remain the same, the epigenetic profile varies remarkably from tissue to tissue, or from organ to organ. We know that DNA modifications are able to store heritable epigenetic information that affects gene expression, and that these are mechanistically linked.

There are several ways that the proteins in compacted DNA can have additional information added as the organism responds to its environment. Methylation, DNA folding pattern remodelling and acetylation are just some of the ways that the RNA network can chemically alter the DNA surface architecture so that a carefully selected combination of genes is either switched 'on' or 'off'. By switching a gene 'on' we mean making it physically exposed and chemically available to be expressed by transcribing it into RNA. Genes that are switched 'off' are those that are either physically inaccessible due to DNA folding and condensation, or have a chemical marker attached to make it impossible for the transcription machinery to run along its length and transcribe it into RNA. Switching a gene off is often referred to as 'silencing'. Methylation is generally associated with silencing. It is a process that can happen very rapidly. Also see: Bird, Adrian (2007). Perceptions of Epigenetics. Nature 447 396-8; Grandjean, Valerie et al (2007). Inheritance of an Epigenetic Mark: The CpG DNA Methyltransferase 1 is required for De Novo establishment of a Complex Pattern of Non-CpG Methylation. PLoS ONE, 11 e1136; and, Choi, Jung Kyoon and Sang Cheol Kim (2007). Environmental Effects on Gene Expression Phenotype Have Regional Biases in the Human Genome. Genetics 175(4) 1607-13.

[70] Eva Jablonka and Marion Lamb have been studying epigenetic data that challenges neo-Darwinian principles for over two decades. For a recent work in which they explore a broad range of inherited phenomena refer to: Jablonka, Eva and Marion Lamb (2007). Evolution in Four Dimensions: Genetic, Epigenetic, Behavioral and Symbolic Variation in the History of Life. Oxford University Press.

[71] Vogt, Gunter et al (2007). Production of different Phenotypes from the Same Genotype in the Same Environment by Developmental Variation. Journal of Experimental Biology 211 510-23.

[72] It was also found that Dolly's telomeres were short when she was born. Using the telomeres as a measure, some believe that Dolly's genetic age at age six was actually twelve, or the same as the sheep from which she was cloned. Some argue that this was the reason for Dolly's early death. Others have argued that the lung disease Dolly suffered from was common among sheep kept indoors. We simply do not know. For a review of the development of deformed foetuses from cloned embryos refer to:

Farin, C.E. et al (2004). Development of foetuses from in vitro-produced and cloned bovine embryos. Journal of American Science 82:E53-E62.

73 Fraga, Mario F. et al (2005). Epigenetic Differences Arise During the Lifetime of Monozygotic Twins. PNAS 102(30) 10604-9.

74 Cropley, J.E. et al (2006). Germ-line epigenetic modification of the murine A^{vy} allele by nutritional supplementation. PNAS 103(46) 17308-12.

75 Kurcharski, R. et al (2008). Nutritional Control of Reproductive Status in Honeybees via DNA Methylation. Science 319(5871) 1827-30.

76 Boucher, B.J. et al (1994). Betel nut (Areca catechu) consumption and the induction of glucose intolerance in adult CD1 mice and their F1 and F2 offspring. Diabetologia, 37 49-55. For a wide review of epigenetics and disease susceptibility see, Jirtle, R.L. and M.K. Skinner (2007). Environmental epigenomics and disease susceptibility. Nature Reviews. Genetics 8(4) 253-62.

77 Chen, Tony H-Hsi et al (2006). Transgenerational effects of betel-quid chewing on the development of the metabolic syndrome in the Keelung Community-based Integrated Screening Program. American Journal of Clinical Nutrition 83(3) 688-92.

78 For an excellent review refer to: Chandler, V. L. & M. Stam (2004). Nature Rev. Genet 5 532-44.

79 Rassoulzadegan, M. et al (2006). RNA-mediated non-mendelian inheritance of an epigenetic change in the mouse. Nature 441 469-74.

80 For a review of transgenerational epigenetic inheritance effects and disease refer to: Whitelaw, Nadia C. and Emma Whitelaw (2008).

Transgenerational epigenetic inheritance in health and disease. Current Opinions in Genetics and Development 18 273-9.

[81] Gosden, Roger and Andrew P. Feinberg (2007). Genetics and Epigenetics – Nature's Pen-and-Pencil Set. The New England Journal of Medicine 356 731-3.

[82] For one of the most convincing studies refer to: Hitchins, M.P. et al (2007). Inheritance of a cancer-associated MLH1 germ-line epimutation. New England Journal of Medicine 356 697-705.

[83] These are some of the major histocompatibility complex (MHC) genes.

[84] Stern, Shay et al (2007).Genome-wide transcriptional plasticity underlies cellular adaptation to novel challenge. Molecular Systems Biology 3(106).

[85] Whatever the answers, these large-scale responses reflect the co-extensive nature of the RNA regulatory network. They also suggest that global responses are based on a set of threshold conditions, and that the processes involved have been designed to meet a wide range of environmental challenges. Some responses are so large, and they involve so many genes that when cells are confronted with the same environmental challenge, the exact transcriptional reprogramming response is non-reproducible in repeated experiments.

[86] Stern, Shay et al (2007).Genome-wide transcriptional plasticity underlies cellular adaptation to novel challenge. Molecular Systems Biology 3(106).

[87] Crick, F. (1984). Memory and molecular turnover. Nature 312(5990) 101.

[88] These early experiments involved tracking DNA methylation changes. It was shown that fear

conditioning in rats is associated with rapid methylation and transcriptional silencing of a memory suppressor gene, and some demethylation for transcriptional activation of a synaptic plasticity gene. These changes were also found to be associated with an increase in mRNA levels. See: Miller, C.A. and J.D. Sweatt (2007). Covalent modification of DNA regulates memory formation. Neuron 53(6) 857-69.

[89] For a recent comprehensive review of the area refer to: Graff, Johannes and Isabelle M. Mansuy (2008). Epigenetic codes and cognition and behaviour. Review Article, Behavioural Brain Research 192 70-87.

[90] McGowan, Patrick et al (2008). Promoter-Wide Hypermethylation of the Ribosomal RNA Gene Promoter in the Suicide Brain. Plos ONE, 3(5) e2985.

[91] Reported to have been said at a PSA meeting in the US, and quoted in, Fox, Maggie (2008). Abuse changes brains of suicide victims. Reuters Science Article, May 6 2008.

[92] Weaver, Ian C.G. et al (2004). Epigenetic programming by maternal behaviour. Nature Neuroscience 7(8) 847-54.

[93] Tsankova, N.M. et al (2006). Sustained hippocampal chromatin regulation in a mouse model of depression and antidepressant action. Nature Neuroscience 9(4) 519-25.

[94] Junko A. et al (2009). Transgenerational Rescue of a Genetic Defect in Long-Term Potentiation and Memory Formation by Juvenile Enrichment. The Journal of Neuroscience, 29(5) 1496-1502.

[95] Elefteriou, Florent et al (2005). Leptin regulation of bone resorption by the sympathetic nervous system and CART. Nature, 434 514-20.

[96] Graff, Johannes and Isabelle M. Mansuy (2008). Epigenetic codes and cognition and behaviour. Review Article, Behavioural Brain Research 192 70-87. (see Table 1, p. 75)

[97] Cannon, Tyronne D. et al (1998). The Genetic Epidemiology of Schizophrenia in A Finnish Twin Cohort: A Population-Based Modelling Study. Archives of General Psychiatry 55(1). Aso see: Kas, M.J.H. et al (2007). Genetics of behavioural domains across the neuropsychiatric spectrum; of mice and men. Feature Review, Molecular Psychiatry 12 324-30.

[98] This has been achieved by virus-mediated remodelling of the DNA folding patterns involved. See: Kumar A. et al (2005). Chromatin remodelling is a key mechanism underlying cocaine-induced plasticity in striatum. Neuron 48(2) 303-14.

[99] Hulpiau, Paco and Frans ven Roy (2009). Molecular evolution of the cadherin superfamily. International Journal of Biochemistry & Cell Biology 41 349-69.

[100] Yap, Alpha S. et al (2007). Making and breaking contacts: the cellular biology of cadherin regulation. Current Opinion in Cell Biology 19 508-14.

[101] Chess A. (2005), Monoallelic expression of protocadherin genes. Nature Genetics 37 120-1; and, Hirayama T. et al (2001). Somatic mutations of synaptic cadherin (CNR family) transcripts in the nervous system. Genes to Cells 6 151-64.

[102] For a comprehensive review of the role of RNAs and the progressive maturation and functional plasticity

of the nervous system in health and disease refer to: Mehler, Mark F. And John S. Mattick (2007). Noncoding RNAs and RNA Editing in Brain Development, Functional Diversification, and Neurological Disease. Physiological Reviews, 87 799-823.

[103] Ader, Robert and Nicholas Cohn (1975). Behaviourally Conditioned Immuno-suppression. Psychosomatic Medicine, 37(4) 333-40.

[104] Gericke G. S. (2008). An integrative view of dynamic genomic elements influencing human brain evolution and individual neurodevelopment. Medical Hypothesis, Elsevier 71 360-73.

[105] A diagram of the layered pangenomic model that includes the main flows of heritable information described in this book is included in the Appendix.

[106] Eldredge, N. and S.J. Gould (1972). Punctuated Equilibria: an Alternative to Phyletic Gradualism. In Schopf, T.M. (ed.), Models in Palaeobiology, Freeman Cooper, pp. 82-115.

[107] The debate continues about exactly how long it took for the 300 species of cichlid fish in Lake Victoria to arise. While the geological and molecular evidence suggests that the Lake Victoria basin dried out around 15,000 years ago, it does not tell us about the types of species that might have remained in the rivers responsible for draining the area. However, Lake Victoria is known to harbour a large number of species that probably originated in situ. For a critical discussion on the origin of the cichlid fish refer to: Nagl, S. et al (2000). The origin and age of haplochromine fishes in Lake Victoria, East Africa. Proceedings of the Royal Society, Biological Sciences 267(1447).

[108] Demouth J.P. et al (2006). The Evolution of Mammalian Gene Families. PLoS ONE 1(1).

[109] Kuhn, T.S. (1962). The Structure of Scientific Revolutions. Chicago, University of Chicago Press.

[110] Richard Dawkins (2005). What We Believe But Cannot Prove. Edited by John Brockman, Pocket Books, p. 9.

[111] Ki Tae Nam et al (2006). Virus-Enabled Synthesis and Assembly of Nanowires for Lithium Ion Battery Electrodes. Science Reports, 312(5775) 885-8.

[112] Aldhous, Peter (2005). Victims of Genetic Discrimination Speak Up. New Scientist, 5 November, 188(2524) p.7.

[113] Hertog, Sofie A.E. de et al (2001). Relation Between Smoking and Skin Cancer. Journal of Clinical Oncology, 19(1) 213-8.

[114] Ferbeyre, Gerado and Scott W. Lowe (2002). The price of tumour suppression? Nature, 3 January 415(6867) 26-7.

[115] S.V. Shinkareva, S.V. et al (2008). Just Using MRI Brain Activation to Identify Cognitive States Associated with Perception of Tools and Dwellings. PLoS ONE 3(1) e1394. doi:10.1371/journal.pone.0001394, January 2, 2008.